Analytics Infrastructure and Risks in the Age of AI

Foundations and Frameworks for Building Enterprise-Scale AI

Arshad Khan

Published by: Khan Consulting & Publishing LLC

Printed in the United States of America

ISBN: 978-0-9660863-9-3

PREFACE

We are living through one of the most consequential technological transitions of our time. Artificial intelligence has shifted from a specialized capability to a defining force across industries, transforming how organizations perceive their environments, make decisions, and create value. Analytics, once grounded in descriptive reporting, has evolved into an adaptive, intelligent ecosystem driven by machine learning, deep learning, generative models, multimodal systems, and autonomous agents. Static insight has given way to dynamic reasoning; human interpretation is increasingly amplified by computational intelligence.

Analytics Infrastructure and Risks in the Age of AI: Foundations and Frameworks for Building Enterprise-Scale AI was written to help individuals and organizations navigate this evolution with clarity and discipline.

Across decades of teaching, consulting, and building analytical systems, I have seen a consistent pattern: AI succeeds not because of algorithms, but because of the foundations beneath them. Organizations struggle when they lack the infrastructure, governance, processes, culture, and risk frameworks required for AI at scale. Many adopt tools without understanding what sustainable, responsible decision-making demands. Others underestimate the ethical, operational, regulatory, and security risks accompanying intelligent systems. Still others try to modernize without building the data and organizational structures that AI depends on.

These recurring struggles reveal a deeper truth: the success or failure of AI is determined long before a model is trained. This book addresses that gap by unifying three domains–analytics, infrastructure, and risk–into a single, coherent framework for enterprise intelligence. Analytics provides insight, infrastructure provides capability, and risk management provides safeguards. Together, they determine whether AI becomes a durable advantage or an unstable experiment.

The chapters of *Analytics Infrastructure and Risks in the Age of AI* trace this full landscape. Early chapters examine the evolution of analytics and the transformational role of AI. Middle chapters outline the architectures, governance structures, and technical environments that support reliable systems, along with the lifecycle practices required to build, deploy, and maintain intelligent models. Later chapters explore organizational, ethical, and regulatory risks, followed by discussions of culture, human-machine collaboration, strategy, and the emerging future of autonomous intelligence.

Although technologies advance rapidly, the principles in this book focus on the durable foundations that make intelligence sustainable: trustworthy data, resilient platforms, disciplined governance, responsible design, and a mature understanding of risk.

My hope is that this book offers structure, clarity, and practical guidance to practitioners, leaders, educators, and students working to integrate advanced intelligence into their organizations. Whether shaping strategy, building systems, developing models, or preparing the next generation of professionals, readers will find a roadmap for navigating complexity with rigor and foresight.

We are entering a new era, one in which intelligence is not an accessory but a foundation. The task before us is to build it wisely, responsibly, and with a clear understanding of its power.

Arshad Khan

TABLE OF CONTENTS

CHAPTER 1

AI as Organizational Infrastructure: A New Foundation for Decision-Making

1.1 Introduction: The Coming of the Intelligence Era

Artificial intelligence has ushered organizations into an era unlike any technological transformation before it. While the industrial revolution mechanized labor and the information revolution digitized processes, the intelligence revolution embeds learning, adaptation, and autonomous reasoning directly into organizational systems. AI is no longer an experimental technology or a back-office enhancement. It is becoming an invisible but indispensable layer of infrastructure, one that shapes decisions, orchestrates processes, and influences outcomes at a scale and speed that was once unimaginable.

Organizations now operate in environments saturated with data, complexity, and rapid change. Traditional tools, structured processes, and even experienced human judgment struggle to keep pace. AI fills this gap not merely by accelerating computation but by enabling systems to analyze patterns, interpret unstructured inputs, predict events, and respond dynamically. In this sense, AI becomes the "cognitive grid" upon which modern institutions operate, similar to how electrical grids supported the industrial era.

This chapter explores what it means to treat AI as organizational infrastructure, including how it transforms operations, reshapes decision-making, alters the nature of competition, and requires new models of governance, ethics, and human collaboration. It establishes the foundation for understanding AI as a core institutional capability rather than a standalone technology.

1.2 The Evolution of Organizational Infrastructure

Every major economic era has been defined by the infrastructure that powered it. Factories relied on steam, electricity, and assembly lines. Information-age enterprises relied on databases, networks, and software platforms. Today, intelligent organizations rely on AI-enabled systems that support perception, prediction, generation, and reasoning.

The shift toward AI-driven infrastructure reflects a deeper transformation in the nature of organizational capability. Traditional infrastructures were deterministic: they functioned the same way every time. AI infrastructures are probabilistic: they evaluate patterns, weigh likelihoods, and generate outputs that depend on context and learning history. This introduces both unprecedented power and inherent uncertainty. The organization must learn to operate in an environment where its core systems do not merely follow rules, they interpret them.

The shift from deterministic to probabilistic infrastructures is as transformative as the shift from manual tools to mechanization. It changes not only the tools an organization uses but the foundation upon which decisions are made.

1.3 Intelligence as a Strategic Resource

The most valuable resource of the modern enterprise is not data alone, but the intelligence derived from it. Intelligence has become the currency of competition, enabling organizations to anticipate change, personalize services, detect anomalies, optimize processes, and create new forms of value.

AI amplifies this resource by converting raw data into continuous insight. Instead of relying on periodic reports or retrospective dashboards, organizations now operate with predictive signals flowing through their systems at all times. AI-driven intelligence informs strategic decisions, automates operational workflows, and identifies risks long before they manifest.

This transition elevates analytics from a supporting function to a strategic capability. Organizations that invest in AI-enabled intelligence systems develop a sustained advantage—an ability to see sooner, decide faster, and act more precisely than their competitors.

1.4 The Expanding Role of AI Across the Enterprise

As AI capabilities mature, their presence within the organization expands from isolated point solutions to enterprise-wide systems. Finance departments use AI to detect irregularities, forecast liquidity needs, and optimize investment strategies. Supply-chain teams rely on machine learning (ML) to predict disruptions, model alternative scenarios, and dynamically route resources. Marketing functions deploy AI to personalize customer journeys, anticipate churn, and recommend targeted engagement strategies.

In operations, AI monitors machine conditions, recognizes defects through computer vision, and orchestrates workflow sequences in real-time. Human resources teams use AI to match candidates to roles, identify skill gaps, and forecast workforce needs. Customer service functions integrate natural language systems that understand intent, resolve inquiries, and escalate issues with contextual awareness.

These applications are not isolated improvements but indicators of a broader shift where AI is becoming the connective tissue of the modern enterprise, supporting knowledge flows and operational decisions across the organization.

1.5 Redefining Decision-Making in Intelligent Organizations

Decision-making has traditionally been driven by experience, intuition, and retrospective analysis. AI introduces a new model in which decisions are informed by continuous, high-frequency signals derived from computational learning.

This model reshapes the very nature of decision-making, where AI expands the decision-making horizon from reactive to proactive, from human-limited perception to machine-augmented awareness, and from static policies to adaptive responses. But this shift also creates new expectations. Leaders must be able to interpret algorithmic outputs, understand uncertainty, evaluate risk, and determine when to rely on AI—and when to override it.

As organizations adopt AI as infrastructure, decision-making becomes a collaborative process in which human judgment and machine insight are intertwined, requiring a governance culture that treats algorithmic recommendations as inputs to be interrogated rather than conclusions to be accepted.

1.6 The Dependency Effect: AI as an Invisible Backbone

As AI systems become embedded in everyday operations, organizations inevitably develop a growing reliance on them. This dependency effect is not inherently negative as every major infrastructure, from electricity to networks, creates deep interdependence. What makes AI different is that it is not a static system. It evolves, learns, and shifts its behavior over time, creating a form of dependency that is dynamic rather than fixed.

This evolving reliance introduces unique forms of risk. AI systems may degrade silently, drift from their original objectives, or become misaligned with current business needs without producing clear warning signals. Unlike traditional infrastructure failures, which are often visible and immediate, AI failures may manifest gradually and only become apparent after significant operational impact.

Managing this dependency requires new forms of risk mitigation. AI infrastructure must be monitored continuously, audited systematically, and maintained with the same rigor applied to critical physical systems, yet with additional safeguards to detect subtle behavioral changes. Organizations must also develop escalation protocols, retraining workflows, and governance processes that account for the system's capacity to change over time.

Recognizing AI as an infrastructural backbone means acknowledging responsibility for its long-term reliability, transparency, and alignment with organizational values. Dependency is inevitable; unmanaged dependency is optional.

1.7 The Importance of Data Governance in an AI-Driven Enterprise

If AI is the intelligence layer, data is the foundation that supports it. AI systems internalize the structure, limitations, and biases of the data on which they are trained. Poor-quality data results in poor-quality intelligence, amplifying noise, skew, and systemic disparities.

Modern data governance must therefore evolve beyond cataloging and compliance. It must encompass representativeness, fairness, metadata richness, lineage transparency, consent management, and ethical sourcing. In traditional analytics, data issues lead to incorrect reports. In AI systems, data issues lead to unintended behaviors, discriminatory outcomes, and deeply embedded systemic risks.

Organizations cannot treat data governance as a procedural requirement. It must be seen as a strategic discipline without which AI-enabled infrastructure cannot function responsibly, ensuring that intelligence generated by AI remains aligned with organizational values and societal expectations.

1.8 Humans and Machines: A New Model of Collaboration

As AI becomes infrastructure, human roles shift from manual execution to supervision, interpretation, and ethical oversight. Humans must understand how AI systems reason, where they may fail, and how to guide them toward desirable outcomes. This creates new expectations for workforce literacy, not only data literacy but AI literacy.

Humans remain essential for context, empathy, accountability, creativity, and the interpretation of ambiguous or morally complex situations. Machines offer speed, scale, and pattern recognition. Together, they form a hybrid intelligence model in which neither replaces the other but each elevates the other's capabilities.

Organizations that fully embrace AI infrastructure must therefore invest in the human side of intelligence, including training, education, governance, and decision support, ensuring that people are empowered not just to use AI but to shape its role in decision-making responsibly.

1.9 The Risks of Treating AI as Infallible

Embracing AI as infrastructure does not mean assuming it is flawless. Intelligent systems are subject to drift, hallucination, misinterpretation, adversarial manipulation, and structural bias. They may perform well under training conditions but degrade when faced with unfamiliar inputs. They may produce results that seem plausible but are fundamentally incorrect.

Overreliance on AI without sufficient oversight leads to automation bias, where human decision-makers defer to algorithmic outputs even when they conflict with reason or evidence. Treating AI as infrastructure requires acknowledging its fragility and building safeguards that prevent blind trust. This includes human-in-the-loop systems, transparent documentation, continuous monitoring, interpretability mechanisms, and clear escalation paths when systems behave unexpectedly.

1.10 AI as a Catalyst for Organizational Transformation

When AI becomes part of the infrastructure, organizations must rethink their processes, culture, and operating model. AI does not simply improve existing workflows; it fundamentally reshapes them. Manual checkpoints become automated, static rules evolve into adaptive policies, and roles centered on execution shift toward oversight, experimentation, and strategic interpretation.

This transformation requires cultural alignment. Organizations that resist change or cling to legacy processes struggle to integrate AI meaningfully. In contrast, those that embrace experimentation, cross-functional collaboration, ethical responsibility, and continuous learning are better positioned to benefit from AI's transformative power. This is because AI infrastructure is not merely technical; it is organizational.

1.11 Conclusion: Entering the Age of Intelligent Infrastructure

Artificial intelligence has become the backbone of modern organizational life. It influences how data is interpreted, how processes unfold, how risks are managed, and how strategic choices are made. It brings unprecedented capabilities but also new vulnerabilities and obligations. Treating AI as infrastructure means recognizing its power, managing its risks, and building systems that combine machine intelligence with human insight.

As the book unfolds, we will explore the evolution of analytics, the foundations of AI-driven systems, the complexities of data governance, the risks inherent in intelligent technologies, and the infrastructure required to support them. Together, these themes form a roadmap for leaders seeking to build and oversee organizations in which intelligence is not just applied, but embedded at the core.

CHAPTER 2

The Evolution of Analytics: From Reporting to Intelligent Systems

2.1 Introduction: A Discipline Transformed

Analytics has evolved through several eras, each shaped by the dominant technologies and organizational needs of its time. What began as a retrospective function used to summarize past events has grown into a dynamic, adaptive capability that informs decisions, optimizes operations, and anticipates change. The accelerating integration of artificial intelligence is now transforming analytics once again. It is not replacing analytics' foundations but elevating them into systems capable of learning, reasoning, and generating insight autonomously.

This chapter explores the evolution of analytics from descriptive reporting to AI-driven intelligence, including how this progression reshapes organizational capabilities, decision-making practices, and the foundations required for intelligent systems. It clarifies how historical approaches continue to influence modern analytical design and adoption.

2.2 The Origins of Analytics: Making Sense of the Past

Analytics began with a simple purpose, to document and understand what had already occurred. Organizations produced periodic reports–monthly, quarterly, annually–to summarize financial performance, operational activity, or customer behavior. These reports were not designed to predict or prescribe actions; their value lay in record-keeping, transparency, and organizational learning.

Early analytics was shaped by scarcity. Computing resources were limited, data was collected manually, and analytical capabilities relied heavily on individual expertise. Most organizations focused on basic questions: What happened? How many? How often? Where? Although rudimentary by today's standards, descriptive analytics established the fundamental principle that decisions should be grounded in evidence rather than intuition alone.

As organizations matured, they sought not just to report on the past but to understand it. This desire set the stage for analytical expansion, eventually opening the door to predictive and prescriptive models that transformed data from a historical record into a strategic asset.

2.3 The Shift to Diagnostic and Predictive Thinking

With more data and better computing power, organizations moved from counting events to explaining them. Diagnostic analytics designed to uncover relationships, correlations, and causal factors emerged, enabling leaders to understand why events occurred and how various elements interacted. These methods improved decision-making by revealing the drivers of outcomes rather than merely the outcomes themselves.

Predictive analytics represented the next major step. Through regression models, time-series forecasts, and statistical algorithms, organizations began to estimate future behavior, demand shifts, risks, and opportunities. Although powerful, these methods were limited by human-designed assumptions. They required analysts to determine which variables mattered and how they related, tasks that demanded both skill and time.

Nevertheless, predictive analytics introduced a new mindset in which organizations looked forward instead of backward. This shift in orientation laid the foundation for AI-driven insight, creating organizational cultures that increasingly valued anticipation, early warning, and strategic foresight.

2.4 The Machine Learning Era: Patterns Beyond Human Perception

Machine learning marked a turning point because it changed the fundamental mechanism of analysis. Instead of relying on predefined formulas and hypotheses, machine learning systems learned patterns directly from data. Models adapted themselves, improved through exposure to examples, and captured relationships that were too complex, nonlinear, or high-dimensional for traditional techniques.

Organizations could now detect fraud more effectively, predict customer churn more accurately, and identify anomalies more reliably. Machine learning introduced adaptability, the ability for analytical systems to evolve as environments changed.

But this era also introduced new requirements, such as larger datasets, cleaner data pipelines, stronger governance, and ongoing monitoring. It became clear that analytics was no longer a one-time activity. It was becoming a living process that would ultimately require new roles, new tools, and entirely new ways of thinking.

2.5 The Rise of Deep Learning and Perceptual Intelligence

Deep learning expanded analytics far beyond structured data. It enabled systems to interpret images, audio, text, video, and sensor signals with unprecedented accuracy. Neural networks learned to represent complexity in hierarchical layers, progressing from simple elements like image edges to entire objects, and from single words to full contextual meaning.

Analytics was no longer confined to spreadsheets or databases. It could read documents, listen to conversations, analyze satellite imagery, and interpret medical scans. Entire industries were disrupted by this shift, including healthcare diagnostics, transportation, security, retail, manufacturing, and entertainment.

Deep learning brought immense capability but also introduced opacity. Models became more powerful yet less interpretable. Their behavior could not always be explained, inspected, or easily verified. Analytics was evolving into a tool that required both trust and verification, setting the stage for the governance challenges that organizations face today.

2.6 Generative AI and Autonomous Insight

The emergence of generative AI transformed analytics again, perhaps more profoundly than any other development. Instead of merely analyzing data, generative models create text, images, insights, summaries, or even reasoning pathways. They synthesize information, interpret ambiguity, and interact conversationally with users.

Analytics is now moving from query-driven to dialogue-driven, from reports to conversational reasoning, and from analyst-generated insights to machine-generated explanations. Generative AI lowers the barrier to entry by allowing non-technical users to access sophisticated insights through natural language.

For the first time, analytics can explain itself, articulating patterns, summarizing trends, and offering hypotheses. But this capability comes with new risks, such as hallucinations, unsupported conclusions, fabricated details, and subtle distortions. Generative analytics must therefore be grounded in robust validation methods and responsible oversight.

2.7 The Changing Role of the Human Analyst

As analytics tools evolved, so too did the responsibilities of the people who use them. In the age of early reporting, analysts were data gatherers and summarizers. As analytics matured, they became model builders, interpreters, and strategists. Today, analysts increasingly serve as supervisors of intelligent systems, evaluating model behavior, interpreting uncertainty, guiding machine reasoning, and ensuring outputs align with organizational goals.

This shift is not a loss of human importance but a redefinition of it. Analytical expertise now lies not in manual data manipulation but in contextual understanding, ethical judgment, domain insight, and the ability to translate complex patterns into meaningful actions. Analysts must understand how AI learns, how it can fail, and how to intervene when it behaves unexpectedly.

Organizations that master AI-enabled analytics are those that cultivate strong human-machine collaboration rather than pursuing automation for its own sake, empowering analysts to act as the critical safeguard that ensures AI systems remain transparent, aligned, and trustworthy.

2.8 The Convergence of Analytics and Decision-Making

The journey from descriptive reporting to AI-driven intelligence has led to a fundamental convergence. Analytics is no longer separate from decision-making; it is woven directly into it. Intelligent systems provide continuous feedback, anticipate emerging issues, and shape actions in real-time. Decisions that once took days or weeks can now be made in minutes or seconds.

This convergence creates new expectations for organizational agility, governance, and strategic alignment. Leaders can no longer rely on periodic snapshots, as they must understand signals that update continuously. They cannot treat analytics as a supporting department; it becomes a strategic foundation. And they must ensure that analytics, now powered by AI, remains fair, transparent, and reliable.

The evolution of analytics is not just technological. It is organizational, cultural, and ethical. It demands new competencies, new accountability structures, and a deeper understanding of how intelligence is produced and used.

2.9 Conclusion: Analytics in an Intelligent World

Analytics has traveled a long distance, from summarizing the past to shaping the future. With the integration of artificial intelligence, it has become a dynamic, learning-driven capability that influences everything from operational workflows to strategic choices. This transformation brings extraordinary potential but also requires thoughtful governance, continuous monitoring, and a balanced partnership between humans and machines.

Organizations that understand this evolution gain a critical advantage. They learn not simply to use analytics, but to build intelligence into the core of their operations. As AI continues to advance, analytics will become increasingly predictive, generative, adaptive, and autonomous–shaping the next era of organizational infrastructure.

CHAPTER 3

What AI Analytics Really Is: Concepts, Capabilities, and Transformations

3.1 Introduction: Understanding the Nature of AI-Enhanced Analytics

Artificial intelligence has transformed analytics from a discipline rooted in structured data and statistical reasoning into one that encompasses perception, prediction, and generative reasoning. What once required analysts to define hypotheses, specify relationships, and manually interpret results can now be executed by systems that learn patterns directly from data, uncover hidden structures, interpret language, and generate insights autonomously. This transformation has broadened the scope of analytics so dramatically that AI analytics now represents a distinct intellectual evolution, a synthesis of traditional analytical thinking with intelligent, adaptive computational models.

Yet the rise of AI analytics does not displace the foundations of the field. Instead, it builds upon them, extending their reach and magnifying their capabilities. Traditional analytics provides transparency, methodological rigor, and interpretability; AI adds scalability, versatility, and adaptive learning. Understanding AI analytics requires appreciating this continuum and recognizing how the two paradigms complement each other.

This chapter explores what defines AI analytics, including the capabilities that distinguish it from traditional analytics and the foundational elements required to build and sustain intelligent systems. It clarifies how intelligence, learning, and adaptation transform analytics from a support function into a strategic capability.

3.2 The Essence of AI Analytics: A Paradigm Shift

At its core, AI analytics refers to the use of artificial intelligence methods to expand the analytical power of organizations, using technologies like machine learning, deep learning, natural language processing, and generative models. Traditional analytics begins with human assumptions: analysts choose variables, define formulas, and generate interpretations. AI analytics begins with data itself, allowing computational systems to infer patterns, relationships, and structures without requiring explicit instruction.

This represents a profound shift in analytical thinking. Instead of instructing systems what to look for, analysts train models that discover what matters. Instead of relying on curated datasets that reflect limited perspectives, AI analytics can incorporate massive volumes of diverse, multimodal data. Instead of producing reports on demand, AI systems generate insights continuously, adjusting to new information as it arrives.

AI analytics, therefore, is not an incremental enhancement but a different way of constructing knowledge. It embodies a model of continuous learning, probabilistic reasoning, and pattern-based inference. It enables organizations to move from understanding what happened to exploring why it happened, what is likely to happen next, and what actions should be taken in response.

3.3 How AI Expands the Boundaries of Traditional Analysis

Traditional analytics excels in structured environments where relationships are stable and well defined. Techniques such as regression analysis, descriptive statistics, and diagnostic comparisons work effectively when data is clean, manageable, and interpretable. These methods remain essential because they offer transparency and accountability, features that AI alone sometimes lacks.

However, traditional analytics is limited by human capacity to specify rules, detect complex interactions, and analyze vast quantities of unstructured information. AI expands these boundaries by enabling models to learn relationships directly from data, capture nonlinearities, and process images, audio, text, and sensor streams with ease. AI systems can adapt to changing conditions, update predictions based on new patterns, and reveal insights that would otherwise remain invisible.

This does not mean traditional analytics becomes obsolete. Rather, the two approaches form a complementary ecosystem. Traditional methods offer interpretability and statistical grounding; AI offers adaptability and depth. Together they provide a more complete analytical toolkit for organizations navigating increasingly complex environments.

3.4 Core Capabilities That Define AI Analytics

AI analytics introduces a suite of capabilities that fundamentally extend what organizations can achieve with data. At the center is predictive intelligence, where machine learning models detect subtle signals in historical behavior and anticipate future outcomes with a level of precision that traditional models rarely match. These capabilities power applications such as demand forecasting, fraud detection, financial risk modeling, and resource optimization, allowing organizations to act earlier and with greater confidence.

A second capability is natural language understanding, which allows AI systems to interpret documents, summarize complex text, extract themes, and engage with users conversationally. This democratizes analytics by enabling employees to ask questions in everyday language, removing the bottleneck of relying on technical specialists to translate business needs into analytical questions.

AI analytics also excels at automated insight discovery, scanning vast datasets to detect anomalies, identify emerging trends, and surface important shifts without manual exploration. This proactive intelligence reduces blind spots and enables faster responses across operations. Relatedly, personalization allows AI to tailor recommendations, decisions, and interventions to individual users, supporting highly customized experiences in domains such as healthcare, education, and consumer engagement.

Finally, AI enables real-time decision-making through streaming analytics and adaptive models that interpret events as they occur. Immediate fraud alerts, dynamic pricing, predictive maintenance, and responsive logistics all stem from this capability. Real-time intelligence closes the gap between insight and action, transforming organizations from reactive observers into proactive operators.

3.5 Data: The Foundation of Intelligent Analytics

Every AI system depends fundamentally on data. But unlike traditional analytics, which can derive meaningful results from carefully curated datasets, AI analytics thrives on diversity, scale, and complexity. Models learn by example, internalizing patterns, structures, and behaviors from the information they receive. As a result, the quality of intelligence produced by AI is inseparable from the quality of the data that feeds it.

Data for AI analytics arrives in many forms, such as structured tables, unstructured text, images, audio, video, logs, time-series signals, and sensor readings. Each type requires specialized preparation. Text must be tokenized and cleaned, images must be normalized, and time-series data must be resampled and synchronized. The process of preparing data for AI remains one of the most demanding and labor-intensive aspects of the analytics lifecycle.

Data quality challenges are magnified in AI systems. Missing values, erroneous labels, sampling bias, representational skew, and historical inequities all leave imprints on model behavior. These imperfections may lead to inaccurate predictions, discriminatory outcomes, or fragile systems that fail in real-world conditions. For this reason, modern organizations must adopt robust data governance practices that extend beyond accuracy and completeness to include fairness, transparency, lineage, and ethical sourcing.

3.6 Algorithms and Models: Engines of Learning

At the heart of AI analytics are the algorithms that learn from data. Machine learning models identify patterns in structured data, while deep learning models can interpret unstructured and high-dimensional information. Natural language models understand linguistic structure, infer meaning, and generate coherent responses. Generative models synthesize text, images, simulations, or synthetic data that support analytical tasks.

Each model type has its own distinct strengths. Classification models help determine whether transactions are fraudulent or whether customers will churn. Regression models predict numerical outcomes such as sales, demand, or price. Clustering algorithms detect segments or behavioral patterns. Neural networks excel at tasks involving vision, language, or complex temporal sequences. Reinforcement learning models optimize decisions through trial and feedback.

However, the power of these models also introduces risks. Complex architectures are often opaque, making the reasoning behind their decisions difficult to explain. Models may overfit, learning noise instead of meaningful patterns. They may generalize poorly to unfamiliar environments or drift as conditions evolve. The sophistication of AI analytics requires equally sophisticated evaluation, monitoring, and governance to ensure that systems remain reliable and aligned with organizational objectives.

3.7 Computational Power: The Engine Beneath the Surface

AI analytics demands far greater computational resources than traditional analytics. Deep learning models require vast numbers of matrix operations, which are efficiently executed only on specialized hardware such as Graphics Processing Units (GPUs) and Tensor Processing Units (TPUs). Cloud platforms allow organizations to access near-unlimited computational power on demand, enabling them to train large models without purchasing their own infrastructure.

Distributed computing frameworks further extend this capability by breaking massive datasets into smaller units processed across many machines. Edge computing brings intelligence directly to devices in the field, reducing latency and enabling real-time analysis at the point of action. This combination of cloud, edge, and distributed computing forms the backbone of modern AI analytics infrastructure.

Without these computational advances, many of the breakthroughs in AI analytics would simply not be possible, and the scalability required for enterprise-level AI adoption would remain entirely out of reach.

3.8 Operational Infrastructure: Deploying and Sustaining Intelligence

Developing an AI model is only one part of the journey; deploying it into real-world operations presents an entirely different set of challenges. AI analytics requires reliable pipelines that deliver updated data, systems for versioning and deploying models, tools for monitoring performance, and mechanisms for detecting drift or anomalies. These processes are collectively known as MLOps, Machine Learning Operations, a discipline that ensures AI systems function predictably and safely after deployment.

MLOps supports continuous learning by automating retraining cycles, validating new versions of models, and ensuring transparency around model changes. It mitigates operational risks by providing visibility into how models behave, how often they are updated, and how they respond to evolving data. Without this operational backbone, AI analytics becomes brittle, untrustworthy, and difficult to scale.

3.9 Human Oversight and Responsible AI

Despite the autonomy that AI analytics provides, human oversight remains indispensable. Humans must define system objectives, evaluate model behavior, question unexpected results, and ensure that algorithmic decisions align with ethical standards and organizational values. Responsible AI practices emphasize fairness, transparency, accountability, privacy preservation, and contextual judgment. These principles are not abstract ideals but practical requirements for any organization deploying AI analytics at scale.

As AI systems increasingly influence hiring, lending, healthcare, public safety, and resource allocation, the stakes grow significantly higher. Ensuring fairness, reliability, and explainability becomes essential not only for organizational compliance but for maintaining public trust. Human oversight acts as a safeguard, preventing automation from amplifying historical inequities, embedding hidden biases, or making high-impact decisions without appropriate scrutiny. In this capacity, humans ensure that AI enhances rather than undermines societal expectations of justice, dignity, and inclusion.

Far from disappearing, the human analyst evolves into a steward of intelligent systems, one who supervises algorithmic behavior, interprets uncertainty, balances risk, and determines when machine recommendations should be adopted, questioned, or overridden. This new model of oversight requires continuous education, cross-disciplinary collaboration, and a governance culture that treats AI as an advisor rather than an unquestionable authority.

3.10 Conclusion: The Foundations of an Intelligent Analytical Ecosystem

AI analytics represents a significant evolution in how organizations generate, interpret, and act on information. By combining the rigor of traditional analytics with the adaptability of machine learning, the perceptual power of deep learning, the accessibility of natural language systems, and the creativity of generative models, AI transforms analytics into a living, learning capability. But this transformation requires strong data foundations, advanced computational resources, operational maturity, and thoughtful human oversight.

As organizations continue their journey into AI-driven intelligence, understanding these foundational elements becomes essential. AI analytics is not defined solely by algorithms or models; it is shaped by data, infrastructure, ethics, governance, and the people who guide and interpret its outputs.

CHAPTER 4

Data Quality: The Bedrock of AI Reliability

4.1 Introduction: Why Data Quality Matters More Than Ever

The rise of artificial intelligence has elevated data quality from an operational concern to a strategic imperative. In traditional analytics, imperfect data might have produced inaccurate reports or misguided interpretations. In AI systems, however, imperfections do far more than distort findings; they alter behavior. AI learns from whatever data it is given, internalizing patterns, biases, omissions, and noise. A flawed dataset becomes a flawed intelligence system, influencing predictions, decisions, and outcomes in ways that can cascade across the organization. Data quality is no longer about clean spreadsheets or timely updates; it is about ensuring that the intelligence layer of the enterprise is built on a foundation that is accurate, representative, ethical, and trustworthy.

Across industries, leaders increasingly recognize that the performance of their AI systems is directly tied to the quality of the data they consume. A model trained on inconsistent labels will make inconsistent decisions. A model trained on biased examples will produce biased recommendations. A model trained on outdated information becomes blind to the present and unprepared for the future. Data quality has always mattered, but in the age of AI, its impact is amplified dramatically. Understanding the principles, risks, and operational realities of data quality is therefore essential for any organization that seeks to build reliable, responsible, and resilient AI systems.

This chapter examines why data quality is the foundational determinant of AI system performance, trustworthiness, and safety. It explores how bias, noise, inconsistency, and incompleteness propagate through intelligent systems and shape downstream outcomes. It also clarifies why poor data quality remains one of the most persistent sources of AI failure in practice.

4.2 The New Meaning of Quality in AI Contexts

Traditionally, data quality was evaluated through a narrow set of attributes: accuracy, completeness, consistency, and timeliness. These remain important, but AI expands the meaning of quality in several profound ways. Where traditional analytics often required data to be structured and standardized, AI must account for unstructured, semi-structured, and multimodal data. Where traditional models could tolerate moderate levels of noise or missing values, AI systems, particularly deep learning models, can be destabilized by subtle irregularities. And where earlier analytics pipelines relied on human judgment to interpret anomalies, AI systems interpret these anomalies as legitimate signals unless explicitly guided otherwise.

Quality in AI environments also extends beyond the dataset itself. The distribution of the data, the representativeness of each segment, the contextual relevance of each observation, and the diversity of examples all shape how well models generalize. A dataset may be accurate yet unrepresentative, complete yet unfair, consistent yet outdated. Quality becomes multidimensional, reflecting not only how correct the data is but how faithfully it reflects the world the model must engage with.

This expansion forces organizations to rethink their approach to data. Quality is not a static condition but an ongoing property that must be protected throughout the data lifecycle, from acquisition and preparation to storage, transformation, and consumption. AI systems cannot filter their own learning inputs; they rely entirely on the quality of what they are given. This dependency changes everything.

4.3 The Fragility of AI Systems to Data Flaws

AI systems are powerful but inherently fragile. Their strength in pattern recognition becomes a weakness when the underlying patterns are distorted, incomplete, or biased. Mislabeled data in supervised learning confuses decision boundaries, irregularities in unsupervised learning create misleading clusters, poor-quality text corrupts natural language models, and even subtle distortions can destabilize deep-learning systems. Small flaws in the data pipeline often cascade into large failures within the model.

A major source of risk emerges when training data fails to represent reality. Models built on narrow demographics, limited behaviors, or incomplete conditions cannot generalize beyond what they have seen, leading to systemic biases that reproduce inequality in outputs. Underrepresented groups or rare events become inaccurately modeled, resulting in discriminatory scoring systems, flawed medical predictions, or misclassified safety incidents. Temporal drift compounds this issue: when historical patterns no longer reflect present conditions, model accuracy degrades silently, turning once-reliable systems into strategic liabilities.

AI's fragility makes data quality not just a technical necessity but a protective barrier. Without rigorous attention to accuracy, representativeness, and ongoing validation, organizations risk deploying models whose failures remain invisible until the consequences are felt at scale.

4.4 Data Diversity, Representativeness, and Fairness

Among the many dimensions of data quality, diversity and representativeness have become especially critical in AI systems. A dataset may contain millions of rows but still fail to reflect the full range of real-world scenarios. An AI system trained primarily on urban environments may struggle in rural ones. A recommendation model trained on majority populations may overlook minority needs. A medical model trained on one region may not generalize to another.

Representativeness is not merely about proportionality; it is about coverage—the ability of a dataset to include the variation, nuance, and edge cases that characterize real-world contexts. Fairness is similarly tied to data diversity. When AI systems produce biased outcomes, the root cause is often traceable to disparities in the training data. Groups that lack representation in the data lack representation in the intelligence derived from it.

Ensuring diversity requires intentional effort. Organizations must actively seek data sources that broaden the context, capture edge cases, mitigate historical biases, and reflect the lived experiences of the populations they aim to serve. Without representativeness, AI systems may function effectively for some users while systematically disadvantaging others.

4.5 The Challenge of Data Integration in AI Pipelines

AI systems rarely depend on a single source of data. They ingest information from transactional systems, sensors, logs, documents, images, and external feeds. Each of these sources has its own format, structure, cadence, and definition of reality. Integrating them into a unified, coherent dataset is one of the most complex challenges in AI analytics.

Differences in units, schemas, time zones, naming conventions, and granularity can create inconsistencies that lead to downstream errors. Joining data incorrectly merges identities. Misaligned timestamps distort sequences. Contradictory definitions across departments introduce silent conflicts. AI systems absorb all of these inconsistencies and treat them as patterns, influencing predictions and decisions in unpredictable ways.

True data quality, therefore, includes not only cleaning individual datasets but harmonizing them across systems. A single error in integration can propagate widely, especially when used in models deployed enterprise-wide. AI does not question the relationships between datasets; it simply learns from whatever relationships it encounters. This places enormous importance on disciplined integration practices.

4.6 Data Preparation: The Transformative Stage

Although often overlooked, data preparation remains the most labor-intensive and intellectually demanding part of the AI lifecycle. It is during this stage that raw, messy, inconsistent information is transformed into a form that models can understand. Preparation involves cleaning, deduplication, normalization, encoding, enrichment, annotation, and feature extraction. Each transformation shapes how the model perceives reality.

Errors at this stage have lasting consequences. A mislabeled training example becomes a misleading rule embedded in the model. A missing value imputed improperly distorts the distribution. A flawed feature-engineering choice emphasizes the wrong pattern. With traditional analytics, preparation errors affect a single report. With AI, they influence every future prediction.

The sophistication of modern AI makes preparation more complex, not less. Multimodal systems require aligned temporal windows across text, sensors, and images. Generative models require careful curation to avoid harmful, toxic, or misleading content. Deep-learning systems require vast amounts of labeled data, often necessitating human annotation or synthetic augmentation. Quality cannot be an afterthought; it must be the central discipline of the AI pipeline.

4.7 Governance and Stewardship: Institutionalizing Data Quality

Data quality depends not only on technical processes but on organizational discipline. Effective governance establishes policies, standards, roles, and accountability mechanisms that ensure data remains reliable across its lifecycle. Data stewards, governance councils, domain owners, and oversight committees collectively uphold these standards. Governance defines how data should be created, validated, stored, transformed, shared, and retired.

In AI environments, governance must also address fairness, privacy, consent, representativeness, and ethical sourcing. Questions that once seemed optional have become essential. Who collected the data, and under what conditions? Who consented to its use? How is sensitive information protected? What risks emerge if the data is misinterpreted? The answers influence the legitimacy and safety of the AI systems that depend on them.

Good governance also acknowledges that data quality is not static. It degrades over time as definitions shift, systems change, and new use cases emerge. Periodic validation, drift monitoring, lineage tracking, and archival processes ensure that the quality of data, and therefore the quality of AI, remains intact.

4.8 The Cost of Poor Data Quality

Organizations often underestimate the cost of poor-quality data, assuming it affects only analysts or reporting teams. In AI systems, the consequences are far broader. Faulty data leads to poor predictions, which lead to misallocated resources, misguided strategies, discriminatory outcomes, and safety risks. A small inconsistency may propagate through models deployed across thousands of users. A mislabeled dataset may influence millions of automated decisions. The cost is not measured only in dollars but in trust, reputation, compliance exposure, and operational stability.

Poor-quality data also undermines confidence in AI. Once users learn that a model behaves inconsistently or unfairly, trust erodes rapidly, even after the underlying issues are resolved. This loss of trust can stall AI adoption, diminish organizational enthusiasm, and ultimately weaken the strategic value of the entire analytics program.

4.9 Conclusion: Quality as the Foundation of Intelligent Systems

Data quality is the bedrock upon which all intelligent systems stand. AI magnifies the strengths of good data and the weaknesses of bad data. It learns whatever patterns it encounters—accurate or flawed, fair or biased, meaningful or incidental. Without high-quality data, no amount of computational power, algorithmic sophistication, or model complexity can produce reliable intelligence.

Organizations that treat data quality as a strategic asset, not an operational burden, position themselves to build AI systems that are trustworthy, effective, and resilient. Those that neglect it invite fragility, bias, and unintended consequences into the heart of their decision-making infrastructure. As AI becomes more deeply embedded in organizational life, data quality becomes not merely a technical requirement but a fundamental principle for responsible and sustainable intelligence.

CHAPTER 5

Data Architecture in the Age of AI

5.1 Introduction: Architecture as the Enabler of Intelligent Systems

Artificial intelligence cannot function without an underlying data architecture designed to support it. No matter how advanced a model may be, it cannot compensate for missing, fragmented, inaccessible, or poorly structured data. In the traditional analytics era, architecture determined how quickly reports could be produced or how efficiently queries could be executed. In the AI era, architecture determines the very capability of a system to learn, adapt, and operate reliably. It is the foundation that governs the flow of information, supports large-scale computation, connects disparate systems, and ensures that the intelligence layer of the organization can develop and evolve.

As organizations accelerate their adoption of AI, they increasingly find that legacy architectures, built for transactional processing or periodic reporting, cannot meet the demands of continuous learning. AI requires high-frequency ingestion, massive storage capacity, rich historical context, real-time streaming, and multimodal information pipelines. It requires architectures that are flexible, scalable, interoperable, and designed with governance and reliability in mind. Data architecture, once viewed as a technical domain, has become a strategic enabler of intelligence and a key determinant of competitive advantage.

This chapter explores how modern data architectures enable scalable, real-time, and multimodal AI systems. It examines the structural foundations that allow data to flow reliably from ingestion through analytics, learning, and operational deployment. It also highlights how architectural choices directly constrain or accelerate AI adoption at scale.

5.2 The Transformation of Data Architecture in the AI Era

Historically, data architecture was relatively stable. Organizations used databases designed for structured information and centralized warehouses optimized for reporting. Extract, transform, and load processes moved data on predictable schedules. Most analytical workloads were performed on static snapshots rather than on continuous flows.

With the rise of AI, this stability has given way to dynamism. Data architecture must now support environments that produce vast amounts of unstructured content, including documents, video, audio, sensor signals, clickstreams, and event logs. The architecture must support continuous ingestion, rapid transformation, and flexible storage models that accommodate both raw and curated data. Instead of rigid schemas, AI systems require architectures that embrace late binding, schema flexibility, and iterative exploration. Instead of daily batches, AI requires near real-time data pipelines that allow models to remain up to date and responsive.

This transformation reflects a broader shift in the role of architecture. It is no longer a passive repository but an active participant in the lifecycle of intelligence. Architecture must support learning, experimentation, deployment, governance, and monitoring, all at the scale and speed demanded by modern AI systems.

5.3 From Warehouses to Lakes to Lakehouses: The Evolution of Storage Paradigms

One of the most significant architectural shifts in recent years has been the evolution from traditional data warehouses to data lakes and, most recently, to lakehouse architectures. Each paradigm emerged to address limitations in the one before it. The traditional warehouse excelled at answering well-defined business questions using structured data, offering clean and organized information but lacking the flexibility to support the unstructured and semi-structured data essential for modern AI workloads.

Data lakes arose in response, providing a scalable environment to store raw, diverse data and enabling broad exploration and experimentation. However, their ungoverned and free-form nature often led to disorder, duplication, and declining trust in the data. The lakehouse emerged as a unifying approach, combining the flexibility and scale of a lake with the management, consistency, and performance characteristics of a warehouse. This hybrid architecture supports raw data exploration alongside curated data pipelines, allowing teams to develop, train, validate, and deploy AI models within a single integrated environment.

These architectural shifts underscore a fundamental principle: data systems must evolve in step with analytical ambition. As organizations move from descriptive reporting to AI-driven insight, their architectures must support greater flexibility, richer data types, and more dynamic learning. The lakehouse model reflects this trajectory, enabling organizations to simultaneously innovate and govern, ensuring that data remains both useful and trustworthy as analytical complexity grows.

5.4 The Role of Real-Time Data in AI Systems

AI systems increasingly depend on real-time data to remain accurate and effective. Fraud detection models need to evaluate transactions as they occur. Predictive maintenance systems must analyze sensor readings in real-time. Personalization engines must respond instantly to user behavior. Supply chain systems must detect disruptions the moment they arise.

Real-time data architectures rely on streaming frameworks that capture, process, and deliver information continuously. Unlike batch pipelines, which move data at scheduled intervals, streaming systems continuously update models and operational dashboards. This allows AI systems to adapt to new patterns and respond to unfolding conditions with minimal delay.

Real-time architecture represents a fundamental shift in organizational responsiveness. Decision cycles accelerate from weeks to days, from days to minutes, and in some environments, from minutes to fractions of a second. Building architectures capable of supporting real-time intelligence requires careful engineering, but the payoff is enormous: organizations gain the ability to act not only intelligently, but instantaneously.

5.5 Integrating Structured and Unstructured Data

AI has expanded the range of data organizations must manage. Structured data is still important, but unstructured content such as text, audio, video, images, transcripts, and social signals has become just as essential. Integrating these two worlds is a defining challenge of modern data architecture.

Structured data is easily queried and analyzed, but it often lacks depth. Unstructured data is rich and expressive, but it is difficult to organize and process. AI systems must merge these two forms to build a complete understanding of events. For example, in customer service, structured data may show the frequency of interactions, while unstructured data from call transcripts reveals sentiment and frustration. In healthcare, structured lab results complement unstructured physician notes. In manufacturing, structured sensor logs combine with video feeds to detect defects.

Architecture plays a crucial role in harmonizing these worlds. It must support multimodal storage formats, flexible indexing, scalable compute layers, and pipelines that can process text, video, and numerical data simultaneously. Only with these capabilities can AI build holistic representations and generate meaningful insights.

5.6 Metadata and Lineage: The Memory System of AI Architecture

As data environments grow more complex, metadata becomes essential. It describes where data came from, how it was transformed, what quality issues it contains, and how it should be used. Metadata provides context, a form of organizational memory, without which AI systems cannot be interpreted or governed responsibly.

Data lineage, a related concept, documents the full journey of data from its origin to its final form. Lineage is crucial for audits, compliance, debugging, reproducibility, and trust. Organizations that lack lineage visibility struggle to understand why models behave a certain way, whether the underlying data has drifted, and what corrective actions are needed.

Strong metadata and lineage practices ensure that architectures support not only the creation of intelligence but also its transparency and accountability. They allow organizations to explain decisions, validate integrity, and maintain the long-term reliability of their AI systems.

5.7 Scalability and Elasticity: Meeting the Demands of AI Workloads

AI systems impose unpredictable and sometimes extreme computational demands. Training large models may require terabytes of data and days of processing; serving real-time predictions may require low-latency, high-throughput architectures. Therefore, scalability and elasticity, the ability to expand and contract resources dynamically, are essential.

Modern architectures rely on cloud-native platforms that can automatically allocate additional storage, compute, or networking capacity as needed. This flexibility allows data teams to experiment, iterate, and innovate without being constrained by physical infrastructure. Elasticity also supports cost optimization, ensuring that organizations pay only for the resources they actively consume.

Architecture must therefore balance performance with adaptability. It must support both experimentation and production, high-volume training and lightweight inference, batch workloads and streaming pipelines. Systems that scale gracefully empower organizations to pursue ambitious AI initiatives with confidence.

5.8 Governance Embedded into Architecture

In the AI era, governance cannot be layered on top of architecture as an afterthought. It must be integrated directly into the design, guiding how data is stored, accessed, transformed, shared, and secured. Architecture must enforce policies for privacy, consent, access control, retention, and quality. It must support automated validation, monitoring, and compliance checks.

Embedded governance ensures that the organization does not depend solely on human diligence to enforce standards. Instead, the architecture itself embodies the rules and protections necessary for responsible AI development. This reduces risk, strengthens trust, and ensures consistency across teams and workflows.

Strong architecture provides not only technical capability but institutional integrity, enabling AI systems to function reliably, ethically, and transparently, while creating a durable framework that scales as data volumes grow and regulatory expectations evolve.

5.9 The Strategic Value of Modern Data Architecture

As organizations adopt AI more deeply, data architecture becomes a strategic asset rather than a technical utility. Architecture shapes how quickly data becomes insight, how reliably models perform, how easily teams collaborate, and how effectively risks are managed. Organizations with modern architectures innovate more rapidly, scale more easily, and deliver intelligence more reliably.

The absence of strong data architecture is equally consequential. Without scalable, flexible, well-governed systems, AI models become fragile, and data becomes inaccessible, fragmented, or inconsistent. Experiments stall, drift accelerates, and trust erodes. In this way, architectural weakness quickly becomes strategic weakness.

Organizations that invest in modern data architecture are not just enhancing their technical capabilities; they are laying the foundation for long-term resilience, adaptability, and competitive differentiation in an AI-centric world.

5.10 Conclusion: Architecture as the Foundation of the Intelligent Enterprise

Data architecture sits at the heart of AI transformation. It determines what data can be collected, how it can be combined, how quickly it can be processed, how reliably it can be governed, and ultimately how effectively intelligence can be created and put into action. In the age of AI, architecture is not simply an enabler. It is the foundation upon which all intelligent systems stand.

Organizations that recognize this truth and invest accordingly position themselves to harness the full potential of AI. They build systems that learn continuously, adapt gracefully, and operate with transparency and reliability. They develop infrastructures that support not only innovation but trust, accountability, and long-term sustainability. In a world defined increasingly by intelligent systems, modern data architecture becomes the backbone of competitive advantage and the cornerstone of responsible AI.

CHAPTER 6

Modern Data Governance for AI Organizations

6.1 Introduction: Governance as the Guardian of Intelligent Systems

As artificial intelligence becomes embedded in organizational infrastructure, data governance has shifted from a procedural responsibility to a strategic mandate. In earlier eras, governance focused on cataloging data, enforcing access rules, and ensuring compliance with regulatory requirements. In the age of AI, governance determines whether intelligent systems behave fairly, reliably, transparently, and ethically. It shapes the quality of insights, the trustworthiness of predictions, the resilience of operational systems, and the credibility of the organization itself.

AI systems do not reason, analyze, or make decisions in isolation. They learn from data, adapt to new information, and amplify whatever patterns they ingest. Without strong governance, these systems inherit risks that may be invisible, systemic, or deeply embedded in organizational history. Governance becomes the structure that ensures data is used responsibly, models are built ethically, and decisions influenced by AI reflect organizational values rather than accidental biases or flawed assumptions. It is the discipline that guards against the unintended consequences of intelligent automation.

This chapter explores the evolution of governance in the AI era, the principles that underpin modern governance frameworks, and the practices organizations must adopt to ensure that data becomes a trustworthy asset and not a liability. It clarifies how governance directly shapes accountability, risk management, and long-term AI sustainability.

6.2 The Changing Role of Governance in the Age of AI

Traditional data governance was fundamentally custodial. It existed to prevent errors, ensure compliance, and create standard definitions. It was primarily reactive, concerned with preserving order and minimizing operational risk. AI disrupts this paradigm by transforming data from a passive resource into an active force—one that feeds predictive engines, shapes automated decisions, and influences customer experiences. As a result, governance must evolve from oversight to orchestration.

In modern environments, governance is not simply about imposing restrictions. It is about enabling innovation safely. Organizations must govern not only data assets but the entire lifecycle of intelligence creation, including the processes through which data is collected, prepared, modeled, deployed, monitored, and retired. Effective governance therefore integrates technical rigor, ethical responsibility, and strategic foresight. It must anticipate how data will be used, how models may evolve, and how the organization will respond when systems behave unexpectedly.

Governance in an AI context becomes multidimensional. It encompasses privacy, fairness, transparency, accountability, lineage, explainability, reliability, and human oversight. It also requires cultural integration, ensuring that people across the organization share a common understanding of how data and AI should be managed. The shift from custodial to strategic governance represents one of the most important evolutions in the modern data landscape.

6.3 Defining Responsibility Across the Data Ecosystem

One of the central challenges in AI governance is the distribution of responsibility. Traditional governance placed most accountability on data stewards, system administrators, and compliance offices. But as AI systems expand across departments, responsibility becomes collective. Data scientists must ensure that their models are built on sound foundations. Engineers must design pipelines that maintain consistency and reliability. Domain experts must validate assumptions, challenge conclusions, and provide context. Leadership must define acceptable use, ethical boundaries, and risk tolerances.

AI governance requires clarity about who is responsible for each stage of the data and model lifecycle. Data owners must ensure accuracy and representativeness. Model developers must test for bias, drift, and robustness. Operational teams must monitor outputs continuously to detect anomalies. Ethics committees must evaluate risks, especially when AI influences high-stakes decisions such as lending, hiring, healthcare, education, or public resource allocation. Responsibility cannot be siloed, because AI systems cross boundaries effortlessly. Governance must reflect this interconnectedness.

6.4 Privacy, Consent, and Ethical Data Use

In an AI-driven environment, the importance of privacy and data rights intensifies. AI systems often require large volumes of personal information, such as behavioral patterns, interaction histories, biometric signals, or unstructured content. As the boundaries between operational data and analytical data blur, organizations face heightened obligations to protect individuals from unwanted surveillance, misuse, or unintended exposure.

Consent becomes more than a legal checkbox; it becomes a matter of ethical legitimacy. Individuals must understand how their data is used, how long it will be retained, and what forms of inference will be derived from it. Sensitive information, once collected under narrow purposes, must not be repurposed without clear justification and explicit permission. The risks of misuse are amplified by the power of AI systems to infer, based on seemingly benign inputs, additional information such as preferences, vulnerabilities, health indicators, or risks.

Governance frameworks must therefore articulate clear data minimization principles, purpose limitations, anonymization practices, and protocols for sensitive data handling. Privacy must be protected proactively rather than reactively. Organizations that treat privacy as part of their ethical identity gain trust; those that treat it as a mere compliance obligation invite systemic risk.

6.5 Ensuring Fairness and Mitigating Bias

AI systems learn from the patterns embedded in their training data. When these patterns reflect historical inequities, misclassifications, or structural discrimination, the system inevitably reproduces and amplifies them. Fairness becomes a governance challenge not because algorithms are inherently flawed, but because data reflects human behavior, institutional history, and societal structures. Bias is rarely intentional; more often it emerges from incomplete representation, skewed sampling, or embedded assumptions that go unexamined.

Modern governance therefore requires rigorous, ongoing evaluation. This includes examining datasets for representational fairness, analyzing model outputs across demographic groups, scrutinizing labeling practices, and identifying missing or distorted information. It also requires transparency around how decisions are made and continuous testing for unintended consequences, since shifts in context or changes in the user population can introduce new forms of bias long after deployment.

Achieving equity in AI demands intentional design, active monitoring, and a governance framework that treats fairness as a foundational requirement rather than an optional enhancement. It is a continuous commitment that ensures AI systems support justice, accountability, and inclusion as they evolve.

6.6 Transparency and Explainability as Cornerstones of Trust

AI systems often operate through complex mechanisms that defy simple explanation. Deep learning models build internal representations that are powerful but opaque. Even simpler machine learning models can behave unpredictably when trained on noisy or unbalanced data. Without transparency, organizations cannot audit decisions, investigate anomalies, or ensure that AI aligns with policy and ethical standards.

Explainability, however, is not merely a technical requirement. It is a social and organizational necessity. Users must understand how AI arrives at conclusions if they are to trust its outputs. Leaders must interpret the strengths and limitations of AI systems when making strategic choices. Regulators increasingly demand explanations for decisions that affect individuals. Governance must therefore define expectations for interpretability, documentation, auditability, and communication.

Modern governance frameworks distinguish between transparency (visibility into how a system is designed) and explainability (the ability to articulate why the system made a specific decision). Both are essential for accountability. An AI system that cannot be explained cannot be governed, while a system that cannot be audited cannot be trusted.

6.7 Lineage, Provenance, and the Traceability of Data

As AI becomes more deeply embedded in operations, the importance of understanding where data comes from and how it evolves becomes paramount. Lineage captures the journey of data from its creation through its transformations, storage layers, feature pipelines, and ultimate use in models. Provenance captures the origin, context, and conditions under which the data was collected. Together, lineage and provenance allow organizations to trace errors back to their source, validate the integrity of datasets, and ensure reproducibility.

Traceability is also essential for compliance. Regulatory frameworks increasingly require organizations to document how decisions were derived, what data contributed to them, and whether proper safeguards were applied. Without lineage, organizations cannot meet these expectations. Governance must therefore ensure that lineage is captured automatically, that it is accurate, and that it remains accessible to auditors and decision-makers. Traceability becomes a foundational component of AI integrity.

6.8 Monitoring, Drift Detection, and Lifecycle Governance

AI systems do not remain static. They evolve as new data enters pipelines, as conditions shift, and as behaviors change. This evolution introduces risks that require continuous oversight. Data drift may introduce new patterns that models were not trained to handle. Concept drift may shift the meaning of variables or relationships. Operational drift may degrade performance due to changes in systems or workflows. Without monitoring, these shifts accumulate silently until the model's reliability is compromised.

Lifecycle governance ensures that AI systems remain aligned with organizational values and requirements from development through deployment, operation, and retirement. It demands periodic validation, performance reviews, fairness assessments, reliability checks, and escalation procedures for anomalies. Monitoring must be proactive rather than reactive, alerting teams to emerging problems before they manifest in real-world failures. Governance ensures that the lifecycle of AI is managed with rigor, transparency, and accountability.

6.9 Embedding Governance into Organizational Culture

Governance succeeds only when it is embraced culturally. If employees view governance as a barrier to innovation, they will circumvent it. If leaders treat governance as an afterthought, teams will ignore it. AI governance must become part of the organizational mindset, shaping how teams collect data, build models, evaluate risks, and deploy systems. This requires education, communication, and reinforcement. It requires leaders to demonstrate that responsible AI is not merely ethical but strategic, and that trust, reliability, and fairness are sources of competitive advantage.

Cultural alignment also ensures cross-functional participation. Governance cannot succeed if data teams work in isolation from legal, compliance, ethics, operations, and domain experts. AI governance must be interdisciplinary because AI systems influence organizational decisions in ways that transcend technical boundaries. The stronger the cultural foundation, the more sustainable the governance program becomes.

6.10 Conclusion: Governance as a Strategic Imperative

Modern data governance is the backbone of trustworthy AI. It ensures that data is collected responsibly, that models are built on sound foundations, and that decisions influenced by AI reflect fairness, accuracy, transparency, and accountability. Governance in the AI era is far more than policy enforcement; it is the discipline that protects organizations from unintended harm, cultivates trust, enables innovation, and ensures long-term sustainability.

Organizations that treat governance as a strategic imperative will build AI systems that are not only powerful but principled. They will navigate regulatory landscapes with confidence, earn stakeholder trust, and use data in ways that elevate rather than endanger their mission. Governance is the framework through which intelligent systems become reliable systems, and through which artificial intelligence becomes aligned with human values.

CHAPTER 7

Data Ethics: Consent, Fairness, Privacy, and Ownership

7.1 Introduction: Ethics at the Core of Data-Driven Intelligence

As artificial intelligence becomes embedded in organizational infrastructure, the ethical dimensions of data move from the margins to the center of strategic decision-making. Data is no longer just an operational byproduct or a raw input for reports; it is the material from which AI systems learn, generalize, and act. The ways in which data is collected, labeled, stored, combined, shared, and interpreted directly shape how intelligent systems behave. When data is treated as a purely technical asset, ethics become an afterthought. When data is treated as an extension of human experience, ethics become inseparable from design.

Ethical questions arise at every stage. Whose data is being collected, and with whose consent? Which groups are represented and which are invisible? How is privacy protected in an era where inference is often more revealing than direct observation? Who benefits from data-driven insight, and who bears the risks? And perhaps most fundamentally, who owns the value created when AI systems learn from the lived experiences embedded in data?

This chapter explores these questions through four interrelated pillars of data ethics: consent, fairness, privacy, and ownership. Together, they define the moral landscape in which AI-enabled organizations must operate. It also examines how ethical failures in any one pillar can undermine trust across the entire system.

7.2 From "Can We Use It?" to "Should We Use It?"

In the early days of digital transformation, organizations approached data questions primarily through a legal lens. If data could be collected, stored, and used within the boundaries of regulatory frameworks, many assumed that it should be. The emphasis was on compliance rather than reflection. AI disrupts this complacency. Because AI systems can infer sensitive characteristics, amplify inequities, and influence high-stakes decisions at scale, the question of what is merely permissible is no longer sufficient. The ethical question becomes: what is appropriate, fair, and respectful in how data is used?

The shift from "can we?" to "should we?" marks a transition from a rule-based mindset to a principle-based one. It encourages organizations to consider the dignity, autonomy, and rights of individuals whose data contributes to AI systems. It prompts a recognition that data is not an abstract commodity but a representation of human lives, relationships, and histories. Ethical data practice therefore requires more than legal compliance; it demands an internal commitment to examining motives, impacts, and unintended consequences.

7.3 Consent in a Complex Data Ecosystem

Consent is a cornerstone of data ethics, yet in practice it is often misunderstood and poorly implemented. Traditional approaches such as opaque terms of service, dense privacy policies, and default opt-in mechanisms fall far short of meaningful agreement, especially in AI-driven environments where data rarely remains confined to its original purpose. Information collected for service delivery may later be repurposed for behavioral prediction, risk scoring, targeted interventions, or model training far removed from the initial context.

Meaningful consent therefore requires more than a checkbox: individuals must genuinely understand what data is collected, how it will be used, what downstream inferences may emerge, and how those uses may evolve over time. This obligates organizations to communicate transparently, obtain renewed consent when purposes shift, and provide accessible mechanisms for withdrawal.

At the same time, there are contexts in which traditional individual consent is impractical or insufficient such as large-scale public datasets, mobility traces, or information derived from collective behavior. In these cases, ethical practice demands additional safeguards, including harm assessments, clear constraints on inference, and independent oversight to prevent misuse. As AI systems generate insights that reach far beyond what individuals can predict or control, consent becomes not just an individual right but a societal responsibility, requiring governance frameworks that balance personal autonomy with collective protection.

7.4 Privacy Beyond Anonymization

Privacy has long been framed as a matter of identifying information such as names, addresses, social security numbers. Often, the solution was anonymization. Data was considered "safe" once direct identifiers were removed or masked. In the age of AI, this assumption no longer holds. Powerful models can re-identify individuals by combining innocuous fragments, can infer sensitive attributes from behavior patterns, and can reconstruct profiles from seemingly unrelated signals. Anonymity is no longer a simple technical transformation; it is a moving target in a world of increasingly capable inference.

Privacy in an AI context must therefore be understood as a broader concept: the ability of individuals to maintain control over how they are seen, categorized, and acted upon by systems. It includes protection against intrusive inference, such as deducing health risks, political leanings, or vulnerabilities from seemingly benign data. It includes limiting secondary uses that people would find surprising or objectionable, even if technically permissible. It includes designing systems that minimize unnecessary collection, retain data only as long as needed, and respect the principle that not everything that can be known should be known.

Organizations must also recognize that privacy is not solely an individual concern. Group privacy becomes relevant when data about communities, neighborhoods, or demographic groups is used to make decisions that affect them collectively. AI systems that segment populations for differential treatment, resource allocation, or targeted messaging raise questions about how privacy, power, and influence intersect.

7.5 Fairness and the Ethics of Representativeness

Fairness in AI is often discussed in terms of algorithms, but its roots lie firmly in data. Models learn from the examples they are given, and those examples encode both the diversity and the inequalities of the world. When some groups are underrepresented in training data, AI systems may make less accurate predictions for them. When historical data reflects discriminatory practices, models may replicate those patterns, reproducing inequities under the guise of objective decision-making.

Fairness therefore begins with representativeness. Ethical data practice requires proactive efforts to ensure that datasets capture the diversity of populations affected by AI systems. This may involve collecting new data, revisiting exclusionary criteria, and challenging assumptions about which cases matter. It also involves recognizing that equal performance across groups is not guaranteed, and that fairness must be monitored over time rather than assumed from the outset.

Beyond representation, fairness involves examining the outcomes AI systems produce in practice. Do they consistently assign higher risk scores to certain communities? Do they decline loan applications for similar applicants at different rates? Do they suggest more burdensome interventions for certain patients? Ethical governance demands that organizations regularly analyze disparate impacts, investigate their causes, and take corrective action. Fairness is not a static property of a model; it is an ongoing commitment to equitable behavior.

7.6 Ownership and the Question of Who Benefits

Ownership of data is one of the most contested ethical questions in modern AI. Traditional perspectives treat data collected by an organization as its asset, to be used, monetized, or shared at its discretion. Yet the individuals whose actions, preferences, and experiences generate that data often receive little control and limited benefit. When AI systems trained on user interactions create new value, such as personalized services, predictive products, or strategic insights, the distribution of that value can become profoundly asymmetrical.

The ethical concept of data ownership asks: who has legitimate claims over the data, and who has the right to decide how it is used? Some argue for stronger notions of personal data sovereignty, in which individuals retain ownership and organizations act as stewards or licensees. Others emphasize collective ownership models, particularly for data originating from communities, public systems, or shared environments. Still others view data less as property and more as a relationship, where rights and responsibilities are negotiated rather than absolute.

Regardless of legal frameworks, ethical practice requires organizations to recognize that value created from data is not entirely self-generated. It is built on contributions, often unrecognized, of users, customers, citizens, and communities. Respecting this reality may involve greater transparency about monetization strategies, mechanisms for individuals to access and port their data, options to opt out of certain uses, and emerging models in which individuals or groups share in the benefits created by AI systems that rely on their data.

7.7 Purpose, Context, and the Risk of Function Creep

Data collected for one purpose may be highly attractive for others. A dataset built to improve service quality can also be tempting for targeted marketing, risk assessment, or surveillance. Over time, the distance between original intent and current use can grow so large that individuals would scarcely recognize the purposes to which their data is being applied. This phenomenon, often called function creep, is one of the most pervasive ethical risks in AI-enabled environments.

Ethical data practice emphasizes purpose limitation and contextual integrity. The idea is simple: the appropriateness of data use depends not only on the data itself but on the context in which it is used. People share information in specific settings with certain expectations; using that same information in a radically different setting may violate those expectations, even if no law is broken. AI intensifies this risk because the same underlying data can be repurposed for increasingly sophisticated forms of prediction and control.

Organizations that take ethics seriously establish clear boundaries around purpose, periodically revisit whether new uses remain aligned with original expectations, and seek new consent or independent oversight when purposes change materially. They recognize that simply because a data use is technically possible, and even strategically attractive, does not mean it is ethically sound.

7.8 Sensitive Data and the Ethics of Inference

Certain types of data have long been recognized as sensitive, such as health records, financial information, biometric identifiers, and details about children, victims of crime, or other vulnerable populations. In the age of AI, however, sensitivity is no longer tied solely to explicit categories. Systems can infer sensitive attributes from non-sensitive inputs by using browsing behavior, mobility patterns, or purchasing habits to estimate health risks, economic instability, or political alignment.

This capacity for inference raises new ethical questions. Should organizations infer attributes that individuals have not chosen to disclose? Should they act on those inferences in ways that affect access to services, pricing, or opportunities? Should inferred vulnerabilities be used to target persuasive messaging or nudge behavior? Where is the line between helpful personalization and exploitative manipulation?

Ethical practice suggests caution in drawing strong conclusions from inferred attributes, humility about their accuracy, and restraint in their application. It also requires clear internal guidelines that distinguish acceptable inference, such as those used to improve safety or accessibility, from uses that encroach on autonomy, dignity, or fairness. Sensitive inferences must be subject to heightened scrutiny, governance, and justification.

7.9 Building Ethical Norms into Data and AI Governance

Ethical data practice cannot be achieved through technical measures alone. It requires institutional structures and cultural commitments that integrate ethics into governance, decision-making, and everyday workflows. Policies must articulate how consent is obtained, how privacy is protected, how fairness is monitored, and how disputes or harms are addressed. Oversight bodies or ethics committees must have genuine authority to question projects, pause deployments, or demand redesign. Data scientists and engineers must be trained not only in modeling techniques but in ethical reasoning.

Crucially, ethics must be seen as integral to quality, not as an obstacle to innovation. An AI system that violates privacy, entrenches bias, or disregards ownership concerns may deliver short-term gains but will ultimately undermine trust and sustainability. Organizations that embed ethics into their data and AI governance frameworks position themselves to innovate responsibly, comply more easily with evolving regulations, and maintain legitimacy with stakeholders.

Ethics also benefits from diversity. Teams that include multiple perspectives are better equipped to anticipate harms, identify blind spots, and design systems that respect different experiences and needs. Ethical governance is not a static checklist but a living process that evolves as technologies, norms, and expectations change.

7.10 Conclusion: Ethics as the Foundation of Trusted Intelligence

Data ethics is not a separate layer added after systems are built. It is the foundation on which trustworthy AI must rest. Consent ensures that data use respects autonomy. Fairness ensures that AI systems do not deepen existing inequities. Privacy ensures that individuals retain control over how they are seen and categorized. Ownership ensures that the value created from data acknowledges and respects the contributions of those who generated it.

As AI becomes the infrastructure of modern organizations, these ethical pillars become as essential as technical performance. Systems that ignore them may function impressively in the short term but will eventually face resistance, regulation, reputational damage, or internal crisis. Systems that are built with ethics at their core have a better chance of earning lasting trust, supporting human well-being, and advancing organizational goals in a way that aligns with societal values.

CHAPTER 8

Machine Learning Foundations for Analytics Leaders

8.1 Introduction: Why Leaders Must Understand Machine Learning

Artificial intelligence in organizations is built largely on the foundation of machine learning. Even when leaders do not design models or write code, their decisions about strategy, investment, governance, and risk depend on a practical understanding of how machine learning works, what it can and cannot do, and where its vulnerabilities lie. Too often, machine learning is treated either as a mysterious black box or as a magical solution to every problem.

Both extremes are dangerous. A lack of understanding can lead to unrealistic expectations, misplaced trust, underestimation of risk, and missed opportunities. A grounded, non-technical grasp of machine learning allows leaders to ask the right questions, set realistic goals, and shape the conditions under which intelligent systems are developed and deployed.

Machine learning is, at its core, a way of learning patterns from data. Instead of writing explicit rules for every scenario, algorithms are allowed to infer those rules by observing examples. This creates systems that can adapt to complexity, handle large volumes of information, and respond to environments that change over time. But the same qualities that make machine learning powerful also make it fragile. If the examples are biased, the learned rules are biased. If the examples are incomplete or noisy, the learned rules are unstable. If the environment shifts, the learned patterns can become obsolete. Understanding these dynamics is essential for anyone responsible for the success of AI initiatives.

This chapter introduces the core concepts of machine learning through a managerial and strategic lens rather than a mathematical one. It equips leaders with the conceptual tools needed to evaluate, guide, and govern intelligent systems effectively. It also clarifies how learning models translate business objectives into adaptive analytical behavior.

8.2 What Machine Learning Really Is

Machine learning is often described as a subset of artificial intelligence focused on enabling computers to learn from data. That description is accurate but incomplete. In practice, machine learning is a discipline that combines mathematics, statistics, computer science, and domain knowledge to approximate relationships between inputs and outputs. Given examples of past behavior, a machine learning system tries to construct a mapping that will perform well on new, unseen situations.

The most familiar form of machine learning is supervised learning, in which each example in the training data has both an input and a known outcome. For instance, a past set of loan applications might include applicant characteristics as inputs and repayment outcomes as labels. The learning process attempts to discover the relationship between these inputs and labels so that the system can predict outcomes for new applicants.

Unsupervised learning, in contrast, has no predefined labels. Instead, it looks for structure, groups, or patterns within the data itself. Reinforcement learning involves an agent interacting with an environment, making decisions, receiving feedback in the form of rewards or penalties, and gradually learning strategies that maximize long-term benefit.

Despite these differences, all forms of machine learning share a common idea: they improve performance by adjusting internal parameters based on data, rather than relying purely on fixed instructions written by humans. This is what distinguishes machine learning from traditional programming. However, it also means that human control shifts from specifying explicit logic to designing the conditions under which learning occurs, including choosing data, objectives, constraints, and evaluation criteria.

8.3 Data as the Central Ingredient of Learning

For leaders, the most important truth about machine learning is that data is not just an input; it is the substance from which intelligence is formed. A model can only learn patterns that exist in the data it sees. If the data only captures a narrow segment of reality, or reflects historical inequities, or contains hidden flaws, the model will faithfully internalize those characteristics. In practice, this means that machine learning does not neutralize bias or noise; it amplifies whatever it inherits.

The learning process typically begins with a dataset divided into subsets for training, validation, and testing. The training subset is used to help the model discover patterns. The validation subset is used to tune settings and prevent overfitting, where the model learns the training data too precisely and fails to generalize. The test subset provides an unbiased check on how well the model performs on unseen data. This separation is crucial, because a model that performs perfectly on the data it has already seen may perform poorly on new, real-world cases.

From a leadership perspective, the quality and representativeness of data deserve as much attention as the sophistication of algorithms. Investments in data collection, cleaning, labeling, and governance are not peripheral technical tasks; they are strategic investments in the reliability of machine learning outcomes. When data is weak, no algorithm, however advanced, can save the system from producing unreliable or unfair results.

8.4 How Models Learn: Error, Optimization, and Generalization

Machine learning models learn by making mistakes and gradually reducing them. During training, a model generates predictions on each example and compares them to the known outcomes. The difference between the prediction and the truth is called error or loss. The learning algorithm then adjusts the model's internal parameters to reduce this error, step by step, across many iterations. Over time, and with properly designed training, the model converges toward a configuration that performs well across the training dataset.

However, good performance on training data is not enough. The true test of a model is how well it generalizes, meaning how accurately it performs on data it has never seen before. Achieving good generalization is a delicate balance. If a model is too simple, it may fail to capture important patterns, leading to underfitting. If it is too complex, it may memorize the training data, capturing noise and quirks rather than underlying relationships. This leads to overfitting and poor performance on new cases.

Leaders do not need to understand the mathematics of optimization algorithms, but they do need to appreciate the trade-offs between complexity and generalization. Requests for increasingly intricate models or additional features may improve performance on historical data but degrade performance in production. Demanding perfect accuracy is often unrealistic and may push teams toward overfitted solutions or ethically questionable shortcuts. A more balanced goal is to seek models that are consistently reliable, transparent enough to be evaluated, and resilient to shifts in the environment.

8.5 Evaluation: Beyond Single-Number Metrics

Evaluating a machine learning model is not as simple as checking a single accuracy score. Different problems require different metrics, and the choice of metric has ethical as well as operational implications. In risk-sensitive domains, such as fraud detection or medical diagnosis, the cost of false positives and false negatives can differ dramatically. A system that incorrectly denies a legitimate transaction may frustrate a customer, while a system that fails to detect fraud may lead to financial loss. In healthcare, a missed positive case may be far more serious than a false alarm.

Effective evaluation requires aligning metrics with real-world impact. This means involving domain experts who understand the consequences of errors, not just data scientists who can compute performance scores. It also means evaluating performance across different segments of the population, not just in aggregate. A model that looks strong on average may perform poorly for particular groups, leading to inequitable outcomes. Leaders should encourage evaluation practices that reflect the full range of ethical and operational requirements, rather than relying on a single metric that hides nuance.

Furthermore, evaluation should not stop at the moment a model is deployed. Continuous monitoring is essential because performance can degrade over time as data distributions change, user behavior evolves, or external conditions shift. A model that was carefully validated last year may be misaligned with reality today. Evaluation, like learning itself, must be ongoing.

8.6 Interpretability, Transparency, and Trust

One of the most challenging aspects of machine learning for leaders is the issue of interpretability. Many of the most powerful models, including deep neural networks, operate through layers of internal computation that do not lend themselves to simple explanation. While they may achieve impressive accuracy, their reasoning is difficult to articulate in human terms. For some applications, this opacity may be acceptable. For others, especially those affecting rights, opportunities, or safety, it is problematic.

Interpretability matters because it underpins trust, accountability, and regulatory compliance. Decision-makers need to know why a model produced a particular recommendation, not merely that it did. Users need to understand the basis of outcomes that affect them, such as being denied a loan or flagged for additional scrutiny. Regulators may require traceability and explanation for high-stakes automated decisions.

There is no single solution to this challenge. Sometimes organizations adopt inherently more interpretable models, even if they are slightly less accurate, because the benefits of transparency outweigh the marginal gain in prediction quality. In other cases, they use post-hoc explanation techniques that approximate how a complex model arrived at a specific decision. In all cases, leaders should be aware that interpretability is not a purely technical consideration but a governance choice. The level of transparency required depends on the stakes involved, the expectations of stakeholders, and the organization's ethical commitments.

8.7 Limitations and Failure Modes of Machine Learning

Machine learning is powerful, but it is not magical. It does not understand context the way humans do, nor does it possess common sense, moral judgment, or the ability to question its own assumptions. It learns statistical patterns, not meaning. When those patterns change, or when it is confronted with cases outside the scope of its training, machine learning can fail in ways that are both surprising and consequential.

Common failure modes include sensitivity to noise, misinterpretation of rare cases, brittleness when confronted with adversarial input, and overconfidence in incorrect predictions. Models may also degrade silently as the world changes, a phenomenon known as drift. Without monitoring, organizations may not notice these failures until they manifest as serious operational issues or public harms. Machine learning systems can also interact with each other in unexpected ways when used as building blocks in larger architectures, creating emergent behaviors no one explicitly designed.

Leaders must therefore approach machine learning with both ambition and humility. It is a tool of remarkable capability, but it is not a substitute for judgment. Its limitations demand safeguards, including human oversight, careful governance, stress testing, and conservative deployment in high-stakes environments. A realistic understanding of what machine learning cannot do is as important as enthusiasm for what it can.

8.8 The Role of Domain Expertise and Human Judgment

Machine learning does not exist in a vacuum. Its success depends heavily on the integration of domain expertise and human judgment. Domain experts know which variables matter, which patterns are plausible, and which relationships would be cause for concern if discovered. They help frame the problem, interpret the results, and identify when the model is behaving in ways that contradict reality or ethical norms.

Human judgment is also essential for deciding when and how to act on model outputs. A recommendation system may suggest an optimal decision from a purely quantitative standpoint, but that decision may conflict with legal constraints, ethical obligations, or long-term strategic considerations. Models cannot see beyond the data and objectives they are given. Humans can weigh broader consequences, understand reputational implications, and consider non-quantifiable values.

For leaders, the challenge is to design workflows where machine learning augments human expertise rather than displacing it. This means clearly defining roles, ensuring that humans understand how to interpret model outputs, and creating channels for feedback that allow the system to improve over time. It also means resisting the temptation to treat model outputs as infallible, especially when they conflict with well-founded human insight.

8.9 Leadership Responsibilities in ML Initiatives

Leadership plays a decisive role in whether machine learning initiatives succeed or fail. Leaders are responsible for setting clear, realistic objectives that connect AI work to business value and societal impact. They must ensure that teams are adequately resourced, not only in terms of tools and talent but in terms of data quality, governance support, and cross-functional collaboration. They must create an environment that encourages responsible experimentation while maintaining appropriate guardrails.

Leaders also shape the ethical posture of machine learning programs. They decide whether fairness is treated as a requirement or a suggestion, whether privacy is managed as a compliance burden or a core value, and whether transparency is pursued where possible or minimized for expediency. They determine whether negative findings, including biased outcomes, performance discrepancies, unexpected side effects, are surfaced and addressed or ignored and downplayed. Machine learning is not just a technical endeavor; it is a reflection of organizational culture and leadership priorities.

Ultimately, leadership responsibility includes knowing when to say no. There are problems that machine learning is not well-suited to solve, situations in which the risks outweigh the benefits, and contexts in which human deliberation is irreplaceable. Saying no requires as much clarity and courage as saying yes.

8.10 Conclusion: A Foundation for Intelligent, Responsible Use of ML

Machine learning has become a central engine of AI within organizations, enabling systems to learn from data, adapt to complexity, and operate at scales that far exceed human capacity. Yet its success depends on much more than algorithms. It depends on the quality and ethics of data, the care with which models are trained and evaluated, the humility with which limitations are acknowledged, and the wisdom with which human judgment is integrated.

For analytics and AI leaders, understanding machine learning at a conceptual level is not a luxury; it is a necessity. It enables informed oversight, meaningful engagement with technical teams, and responsible decision-making about where and how AI systems should be deployed. It helps ensure that machine learning is used not just to optimize metrics, but to advance organizational goals in a way that aligns with fairness, privacy, accountability, and long-term trust.

CHAPTER 9

Deep Learning, Perception, and Multimodal Intelligence

9.1 Introduction: The Leap from Learning to Perception

Deep learning marked the moment when machines moved from analyzing structured data to interpreting the world in ways that resemble human perception. Unlike traditional machine learning, which depends heavily on manually engineered features and domain-specific inputs, deep learning systems learn directly from raw data such as images, audio, text, and sensor signals. This shift allowed machines to recognize objects, interpret speech, understand language, detect anomalies, and analyze patterns that were previously accessible only through human intuition.

For organizations, deep learning opened new doors, enabling them to analyze unstructured information at scale, automate perceptual tasks, and uncover insights that were once out of reach.

This chapter explores how deep learning works, how it reshapes perception and representation, how multimodal intelligence is emerging, and how these developments redefine the future of analytics. It also explains why these capabilities mark a fundamental shift in how machines interpret complex real-world information.

9.2 Neural Networks: The Architecture of Representation

At the core of deep learning lies the neural network, a multilayered computational structure capable of transforming raw input into increasingly abstract forms. Although inspired loosely by biological neurons, modern neural networks are mathematical engines that learn representations by adjusting internal weights through exposure to data. Each layer in a network extracts progressively richer patterns. Early layers detect simple signals such as edges or tones, intermediate layers identify shapes or phonemes, and deeper layers construct conceptual meaning from these building blocks. Through this hierarchy, networks do not merely process data. They internalize its structure, forming representations that capture relationships far too intricate for manual feature engineering.

It is important to note, however, that the relationship between learning and memorization is subtle. Deep networks can generalize well, discovering patterns that apply beyond the training dataset, but under certain conditions they may also memorize large portions of their inputs. What makes them powerful is their ability to do both simultaneously, learning representations that reflect the essential patterns of a domain while absorbing enough detail to preserve performance on rare or nuanced cases. This balance between abstraction and memorization underpins the versatility of deep learning and explains why neural networks excel at tasks involving perception, language, and high-dimensional signals.

9.3 Convolutional Models and the Birth of Computer Vision

The emergence of convolutional neural networks (CNNs) transformed the field of computer vision by enabling machines to interpret images through layered, hierarchical understanding. Convolutional models scan small regions of an image, detect edges and textures, and integrate these features into full representations of objects and scenes. This architecture proved exceptionally effective, quickly becoming the foundation for image classification, object detection, facial recognition, quality inspection, and medical imaging analysis. With CNNs, visual data, long considered too unstructured for analytical use, suddenly became accessible and interpretable at scale.

As organizations adopted computer vision technologies in operational settings, often referred to as machine vision, they discovered that visual intelligence could be embedded directly into operational workflows. Manufacturing lines gained automated defect detection, healthcare systems improved diagnostic accuracy, logistics operations used real-time image analysis to monitor conditions, and safety systems leveraged vision models for continuous oversight. Machine vision thus became a powerful extension of analytics, enabling organizations to extract meaning from images just as they had long done with numerical and textual data, elevating visual information into a strategic asset for decision-making.

9.4 Sequence Models: Understanding Language, Sound, and Time

While convolutional models transformed vision, sequence models revolutionized the interpretation of text, speech, and time-sensitive data. Early architectures such as recurrent neural networks (RNNs), long short-term memory (LSTM) networks, and gated recurrent units (GRUs) introduced the idea that machines must process information sequentially, carrying forward context and memory from one step to the next. This shift enabled computers to read sentences as meaningful sequences rather than isolated tokens, recognize speech with far greater accuracy, and analyze evolving signals such as sensor streams, financial markets, or customer activity patterns. These models helped machines interpret sound waves, understand sentiment, and spot temporal anomalies in system logs or equipment data, which are capabilities that traditional statistical methods struggled to capture.

Although transformers eventually surpassed these earlier architectures, the conceptual breakthrough of sequence modeling remains foundational. It established the principle that data must be understood not as static snapshots but as unfolding narratives with past, present, and future implications. This perspective underpins modern natural language processing, multimodal AI, predictive maintenance, real-time analytics, and any domain where meaning emerges through evolution over time. Sequence models taught machines to follow the rhythm, structure, and progression of human and environmental signals. In doing so, they laid the intellectual groundwork for the next generation of intelligent systems.

9.5 The Transformer Revolution and Foundation Models

The invention of the transformer architecture reshaped deep learning by enabling models to analyze relationships between all elements of an input simultaneously rather than processing sequences step-by-step. This global attention mechanism allows transformers to capture long-range dependencies, subtle patterns, and layered meaning with remarkable precision. Just as important, transformers scale efficiently by training on massive datasets and leveraging parallel computation at levels that earlier architectures could not approach. This scalability unlocked dramatic advances in natural language processing, vision, audio modeling, and multimodal reasoning.

These breakthroughs gave rise to foundation models: large, pre-trained networks that learn broad, general-purpose capabilities transferable across thousands of tasks. Instead of being built for a single purpose, these models can be adapted through prompting, fine-tuning, or contextual instructions, enabling them to summarize text, write code, classify images, analyze audio, and reason across domains. The transformer era introduced machines that generalize rather than specialize. They mark a fundamental shift toward flexible, adaptive intelligence—systems that serve as platforms for innovation rather than isolated tools.

9.6 The Expansion of Perception: Vision, Speech, and Sensor Intelligence

Deep learning expanded the range of data organizations can analyze by enabling machines to perceive the world across multiple sensory channels. Information once considered too unstructured, such as images, audio recordings, environmental signals, and continuous sensor streams, became rich sources of intelligence directly integrated into analytical workflows. Through this shift, perception evolved from a uniquely human capability into a scalable computational function, allowing AI systems to observe and interpret complex real-world environments in ways traditional analytics could not.

Computer vision and audio intelligence matured rapidly under deep learning. Vision models learned to detect manufacturing defects, classify medical scans, monitor infrastructure via drone or satellite imagery, and evaluate logistics conditions in real-time. In parallel, speech and audio models achieved accurate transcription, contextual understanding, sentiment analysis, and early detection of mechanical failures through acoustic signatures. What were once opaque or noisy data sources, visual and auditory signals, became reliable inputs for diagnostics, decision support, and operational monitoring across industries.

Sensor intelligence completes this perceptual expansion. Modern factories, vehicles, buildings, supply chains, and digital ecosystems generate continuous streams of readings that reflect movement, vibration, temperature, pressure, and environmental change. Deep learning models interpret these sequences to identify anomalies, degradation, inefficiencies, or unsafe conditions before they escalate. Combined, vision, speech, and sensor intelligence give organizations a holistic, real-time understanding of their operations and risks, which moves analytics beyond numerical reports to encompass the full spectrum of perceptual information embedded in the world around them.

9.7 Multimodal Intelligence: Integrating Perception and Language

The frontier of deep learning is multimodal intelligence, systems capable of interpreting images, text, audio, charts, video, and sensor streams as pieces of a unified context. Unlike earlier models confined to a single data type, multimodal systems understand relationships across modalities. They can read a chart and explain the trend, analyze an image and answer questions about its content, watch a video and summarize events, or interpret a document that contains embedded visuals and numerical tables. This integration brings machine understanding closer to human cognition, where meaning emerges from synthesizing multiple sensory inputs rather than examining each in isolation. As a result, multimodal AI blurs traditional boundaries between structured and unstructured data, enabling richer insights that reflect the complexity of real-world information.

For organizations, multimodal intelligence transforms the analytics experience into something far more natural, intuitive, and interactive. Decision-makers can, all within a single analytical workflow, ask questions about charts, request explanations of images, or explore insights that intertwine numerical patterns with visual or textual evidence. This fluid interaction mirrors how humans naturally reason: by combining what they see, read, hear, and infer. Multimodal systems therefore elevate analytics from a series of disconnected tasks to a seamless, context-aware process that adapts to how people think and work. In doing so, they redefine the future of decision support and bring AI closer to a truly comprehensive understanding of the environments it is tasked to interpret.

9.8 Transfer Learning and the Adaptation of Expertise

Transfer learning has become one of the most valuable techniques in modern AI because it allows organizations to leverage the knowledge embedded in large, pre-trained models and adapt it to specialized applications. Instead of training a model from scratch, a process that often requires extensive datasets and formidable computational resources, teams can begin with a model that has already learned general patterns from millions of examples. By fine-tuning this model on a smaller, domain-specific dataset, organizations achieve strong performance with far fewer resources and much shorter development cycles. This approach enables a financial institution to adapt a language model to legal contracts, a hospital to tailor a vision model to medical imaging, or a manufacturer to tune a vision system for highly specific defects.

While transfer learning dramatically accelerates innovation, it also introduces new responsibilities. A pre-trained model brings with it the data biases, assumptions, and blind spots of the environment in which it was originally trained. These inherited characteristics may persist even after fine-tuning, potentially reproducing inequities or producing unexpected behavior in specialized contexts. The efficiency of transfer learning therefore must be matched with careful evaluation, transparency about origins, and governance over how foundational models are selected and adapted. Without this scrutiny, organizations risk building systems that inherit capabilities they do not fully understand and biases they did not anticipate.

9.9 Strengths, Limitations, and the Fragility of Deep Systems

Deep learning systems achieve remarkable accuracy, but their power is accompanied by inherent fragility. They perform best when new inputs resemble the patterns they encountered during training. When faced with unfamiliar or out-of-distribution examples, their predictions can deteriorate rapidly and often without obvious warning. Models trained on bright, controlled imagery may fail in low-light conditions, and speech systems trained on dominant dialects may struggle with linguistic variation. Deep models can also be manipulated through adversarial perturbations, subtle modifications to an input that appear insignificant to humans but lead the model to confidently misclassify what it sees or hears. These vulnerabilities arise because deep learning captures correlations rather than causal understanding; it recognizes patterns extremely well but does not reason about why those patterns occur.

Fragility also emerges over time as real-world environments shift. Patterns in consumer behavior, economic signals, physical environments, or sensor conditions evolve, creating data drift that gradually misaligns the model's learned representations with reality. A system may continue to produce outputs with high confidence even as its accuracy erodes. This combination of silent degradation, sensitivity to distributional changes, and vulnerability to manipulation underscores the need for rigorous validation, stress testing, interpretability tools, and continuous monitoring. Deep learning amplifies capability, but it also amplifies risk; organizations must treat these systems not as fully autonomous intelligence but as components requiring ongoing stewardship and oversight.

9.10 Deep Learning Within Modern Intelligent Workflows

Deep learning has fundamentally reshaped analytics workflows by extending intelligence far beyond traditional structured datasets. Modern analytical pipelines now integrate visual data, audio recordings, text documents, sensor streams, and real-time signals, which are data types that once sat outside the reach of conventional analytics. Feature extraction, historically the most labor-intensive and specialized step of machine learning, increasingly occurs automatically as deep models learn rich, multilayered representations from raw inputs. This shift allows workflows to incorporate explanations, contextual reasoning, anomaly detection, and visual evidence generated by systems that unify perception and inference. Interfaces are also evolving and, rather than relying solely on dashboards, decision-makers can converse directly with intelligent systems through natural language, asking questions, refining insights, and receiving synthesized interpretations drawn from multiple modalities.

As deep learning integrates more tightly into operational processes, analytics becomes perceptual, adaptive, and continuous. Intelligent workflows monitor environments, recognize emerging patterns, and generate context-aware insights in real-time, giving organizations a dynamic understanding of the present rather than a static snapshot of the past. Decision-makers are supported with narrative summaries, multimodal evidence, and predictive signals that reflect the complexity of real-world conditions. In this sense, deep learning does more than enhance analytics. It transforms it into an interactive, situationally aware partner that augments human judgment across the entire decision-making lifecycle.

9.11 Conclusion: The Foundation for Intelligent Systems

Deep learning established the perceptual foundation for modern AI systems. It gave machines the ability to interpret the world through sight, sound, language, and high-dimensional signals. It enabled representation learning, transfer learning, and multimodal intelligence, which are capabilities that now underpin generative AI and cognitive analytics. Deep learning is not the endpoint but the platform upon which the next stage of intelligence is being built.

CHAPTER 10

Generative AI and the Rise of Autonomous Insight

10.1 Introduction: A New Creative and Cognitive Frontier

Generative AI marks a profound shift in how organizations create, interpret, and interact with information. For decades, analytics and machine learning focused primarily on identifying patterns within existing data. Generative AI extends that paradigm by enabling machines not only to analyze the world but to create new representations of it. Large language models generate fluent text, image models create detailed visuals, and multimodal systems synthesize insight across language, imagery, sound, and data.

This creative capacity transforms analytics from a process of extraction to a process of generation. It allows systems to summarize documents, explain complex models, produce hypotheses, design experiments, and even propose decisions expressed in natural language. Instead of waiting for humans to ask the right questions, generative systems can surface insights proactively. The rise of generative AI marks the beginning of autonomous insight, where machines serve as partners in thought, extending human cognition rather than merely automating mechanical tasks.

This chapter explores how generative models transform analytics from predictive tools into systems capable of reasoning, simulation, content generation, and autonomous insight. It examines both the strategic potential and the new risks introduced by synthetic intelligence. It also clarifies how synthetic outputs reshape decision-making, experimentation, and innovation across organizations.

10.2 The Foundations of Generative Intelligence

Generative AI is built on models that learn the statistical structure of the data they encounter. Rather than applying fixed, rule-based logic, these systems internalize patterns across language, images, audio, and other modalities. This enables them to produce new outputs that resemble, but do not duplicate, their training data. At the center of this capability is the transformer architecture. By using attention mechanisms to analyze relationships across all parts of an input simultaneously, transformers capture nuance, context, and dependency with a depth no prior architecture achieved. This global contextual awareness allows them to learn representations that generalize across tasks rather than serving a single, narrow purpose.

When trained on massive datasets, transformer-based models develop broad competencies that extend far beyond prediction. They can answer complex questions, summarize lengthy reports, translate languages, write code, analyze charts, and generate analytical narratives that adapt to a user's intent. Their flexibility emerges not from explicit programming but from the diversity and richness of the patterns they observe during training. This gives generative models a new kind of *synthetic generality*, a wide-ranging, adaptive capability that marks a fundamental shift in how intelligence can be produced and applied in analytical environments.

10.3 From Predictive Models to Generative Reasoning

Traditional predictive analytics focuses on probabilities, classifications, and numerical forecasts. Generative AI extends beyond this by producing coherent sequences that approximate human reasoning, through text, explanations, recommendations, and structured content. A generative model can interpret charts, articulate trends, summarize reports, propose correlations, or analyze operational logs to suggest likely causes of failure. These abilities are grounded not in symbolic logic but in pattern-based reasoning. The model draws on its internalized patterns to generate responses that resemble how humans might think through a problem. While generative models do not *understand* information in the human sense, their ability to mimic reasoning enables them to assist with complex cognitive tasks that once required expert interpretation.

This shift transforms analytics from a tool for prediction into a tool for explanation and ideation. The machine becomes capable of producing narratives, hypotheses, and context-aware interpretations that complement human analysis. However, generative reasoning is also delicate. It depends on proper model alignment, grounding in factual data, and safeguards that prevent hallucination or misuse. When appropriately governed, generative AI becomes a companion to human judgment, expanding the analyst's capacity to explore possibilities, understand uncertainty, and communicate insights with unprecedented fluency.

10.4 The Emergence of Multimodal Generative Systems

The generative revolution has expanded beyond text into multimodal intelligence, which are systems that integrate vision, language, audio, structured data, and sensor information into unified models capable of producing cross-modal insights. A multimodal generative system can look at an image and provide a verbal explanation, read a chart and produce a narrative analysis, interpret a document and reference embedded diagrams, or observe a video and summarize events in context. By connecting modalities that were once processed independently, these models begin to mirror the integrative nature of human perception and communication, understanding information as a blended whole rather than as isolated data types.

For organizations, multimodal generative systems fundamentally reshape the analytical landscape. Analytics is no longer limited to numerical tables or textual reports, as images, recordings, diagrams, logs, and sensor streams become equally analyzable and generative. Decision-makers can interact with these systems conversationally, receiving rich, contextual insights synthesized from diverse formats. Multimodal intelligence unifies information sources that previously required separate tools and specialists, creating a single interface through which users can explore complex environments, uncover relationships, and make decisions informed by the full spectrum of organizational data.

10.5 Generative AI as a Cognitive Assistant

Generative AI represents a major shift in how individuals interact with information. Acting as a cognitive assistant, it helps users interpret data, refine ideas, analyze alternatives, and navigate complex decision-making processes. A leader reviewing financial or operational reports can ask the model to highlight anomalies, compare scenarios, or estimate potential risks. Analysts can instruct the model to propose features, write code, generate test cases, or summarize lengthy outputs. Customer-support teams can use generative AI to condense thousands of conversations and surface the root causes of dissatisfaction.

The defining characteristic of generative cognitive assistance is adaptability. The system adjusts to the user's level of expertise, communicates in natural language, and works interactively through iterative dialogue. It accelerates thought processes by reducing the time needed to explore possibilities, test assumptions, or produce drafts. Unlike traditional analytics tools, which require users to master interfaces or modeling techniques, generative models lower the barrier to entry, enabling non-technical users to engage meaningfully with complex information. This personalized support transforms generative AI into an always-available analytical partner that enhances human reasoning without replacing judgment.

10.6 The Challenge of Hallucination and Reliability

Despite its strengths, generative AI carries inherent limitations. The most notable is hallucination, the production of outputs that are fluent, detailed, and persuasive yet factually incorrect. These errors occur because generative models optimize for coherence and likelihood rather than truth. They assemble responses based on statistical patterns rather than validated facts, meaning that a fabricated trend, an invented numerical value, or an incorrect explanation may be delivered with full confidence. In analytical settings, such mistakes can mislead decision-makers, distort insights, or propagate misinformation through downstream workflows if not properly checked.

Ensuring reliability requires grounding generative systems in factual data sources, implementing verification and cross-checking mechanisms, and embedding workflows where human judgment remains central. Generative AI should be treated as a hypothesis generator—capable of rapidly proposing ideas, summarizing complex information, or revealing alternative interpretations—but not as an authoritative source of truth. Its strengths lie in creativity, exploration, and speed; its weaknesses lie in its tendency to overgeneralize or invent details. Responsible use depends on safeguards that maintain trust and accuracy without constraining the creative potential that makes generative AI so valuable.

10.7 Ethical, Social, and Legal Implications of Generative Intelligence

As generative systems become more capable, they bring forward a range of ethical, social, and legal challenges. Because they learn from large-scale datasets containing societal biases, they may inadvertently reproduce or amplify those biases in their outputs. They can also create highly realistic but false content, such as deepfakes, fabricated documents, or misleading narratives that blur the boundary between truth and fiction. In sensitive domains such as hiring, healthcare, finance, or public safety, these risks carry significant consequences. Organizations must also consider issues of intellectual property, as generative models may reproduce stylistic elements or patterns from copyrighted material, raising questions about attribution, ownership, and fair use.

From a regulatory perspective, generative AI challenges traditional accountability structures, as outputs emerge from statistical inference rather than explicit programming. Organizations using these systems must adopt clear frameworks for transparency, auditability, and responsible use. This includes determining when human review is mandatory, how AI-generated content should be labeled, what safeguards prevent misuse, and how harms are identified and mitigated. The potential benefits of generative AI are substantial, but realizing them safely requires governance practices that ensure the technology is deployed ethically, equitably, and with respect for individual and societal rights.

10.8 The Role of Synthetic Data in AI-Driven Analytics

Generative models have introduced the powerful capability to create *synthetic data*, artificial datasets designed to mimic the statistical properties of real-world data without revealing sensitive or identifiable information. Synthetic data can augment training sets, mitigate imbalance, simulate rare or dangerous events, and enable innovation in fields such as healthcare, finance, manufacturing, and cybersecurity. It allows teams to test models under extreme conditions, explore hypothetical scenarios, and conduct research that would otherwise be limited by privacy constraints or data scarcity. When used appropriately, synthetic data helps organizations innovate while protecting individuals' rights and reducing reliance on sensitive datasets.

However, synthetic data inherits the strengths and weaknesses of the models that generate it. If real data contains inequities, synthetic data may reinforce or amplify those biases. If the underlying model is trained on incomplete, skewed, or outdated information, its synthetic outputs may distort the reality they are meant to approximate. Poor-quality synthetic data can lead to misguided conclusions, overfitted models, or unexpected failures in production. For this reason, synthetic data requires robust governance, which can ensure transparency about how it is generated, rigorous evaluation of its quality and representativeness, and safeguards to prevent misuse. When properly managed, synthetic data becomes a strategic asset. However, when mismanaged, it undermines the very reliability organizations seek to improve.

10.9 Human–AI Co-Creation and the New Workflow of Insight

As generative AI becomes integrated across organizations, the workflow of analytics shifts from linear production to collaborative co-creation. Instead of relying solely on analysts to produce outputs manually, teams increasingly work in partnership with models that generate drafts, propose hypotheses, surface insights, and assemble content. Humans shift into roles as editors, evaluators, and contextual decision-makers, refining and validating the machine's contributions. This evolution expands analytical capacity, enabling organizations to process more information, iterate faster, and explore a wider range of possibilities.

Co-creation also reshapes organizational expectations. Strategic leaders use generative AI to explore scenarios, stress-test narratives, or examine the implications of decisions. Subject-matter experts collaborate with the model to extend their reach, producing explanations, summaries, documentation, or prototypes with greater speed. Cross-functional teams rely on generative tools to bridge communication barriers, as the model can translate technical concepts into plain language or synthesize input from different stakeholders. This partnership requires a new form of fluency, an ability to guide models with clarity, question outputs with rigor, and integrate human insight with machine-generated material. When implemented well, co-creation enhances both productivity and accountability, making analytics a more inclusive and interactive organizational process.

10.10 Conclusion: Toward the Era of Autonomous Insight

Generative AI represents the next major stage in the evolution of analytics. It extends analytical capability beyond prediction and classification, enabling systems to create narratives, interpret complex information, and offer insight in human language. It unifies perception, reasoning, and creation into a single capability that augments human thought. Yet its strength must be matched with caution. Generative models are powerful but imperfect, innovative but fallible, creative but occasionally misleading. The path forward requires a balance between automation and oversight, creativity and control, speed and scrutiny. As organizations adopt generative systems, they enter an era where insight is no longer limited by human bandwidth but must be guided by human judgment.

CHAPTER 11

Natural Language Analytics: Conversational Interfaces and Insight Generation

11.1 Introduction: When Analytics Learns to Speak

Natural language analytics represents one of the most transformative shifts in the evolution of intelligent systems. For decades, analytics required users to write queries, navigate dashboards, interpret charts, and understand data structures. Natural language analytics removes these barriers by allowing individuals to interact with data conversationally. A user can ask questions in plain language, request explanations in narrative form, and explore datasets through dialogue rather than technical manipulation. This shift democratizes analytics, making insight accessible to employees regardless of their technical background. It also creates a more fluid relationship between humans and data, where the system acts as an advisor, translator, and interpreter rather than a passive repository of reports.

Natural language analytics did not emerge in isolation. It builds directly on the advances in deep learning, generative models, and multimodal intelligence. These underlying technologies enable systems to understand linguistic nuance, produce coherent explanations, interpret visual information, and connect natural language queries to structured data sources. As a result, analytics becomes both more powerful and more intuitive. The boundary between data and decision-making narrows, replaced by a conversational interface that brings insight to the point of need.

This chapter examines how natural language interfaces dissolve technical barriers between humans and data. It explores how conversational analytics reshapes decision-making, accessibility, and the role of analysts in intelligent organizations. It also highlights how language becomes the primary interface for interacting with intelligent systems.

11.2 How Machines Understand Language

At the core of natural language analytics lies the ability of models to interpret human language with high levels of fluency and contextual awareness. Modern models do not understand language the way humans do; rather, they learn patterns through exposure to vast amounts of written text. Through deep learning and transformer architectures, they develop internal representations of grammar, meaning, relationships, and reasoning patterns. These representations allow them to map natural language questions, such as "Why did revenue decline last quarter?", to relevant analytical tasks, such as identifying trends, isolating anomalies, or generating comparisons.

This linguistic capability includes not only understanding questions but also interpreting context. A system remembers the conversational thread, understands follow-up clarifications, and adjusts its responses accordingly. This conversational coherence transforms analytics from a series of one-way requests into a dynamic exchange. The system becomes capable of asking clarifying questions, suggesting alternative interpretations, or highlighting uncertainties that require human judgment. In this way, language becomes the interface through which insight flows.

11.3 Conversational Querying and the End of Rigid Syntax

Traditional analytics required users to write SQL queries, manipulate filters, or design dashboard views. Conversational querying removes this dependence on technical syntax. A user can ask, "Show me the top-performing regions," or "Compare this year's demand to the previous three years," or "Explain why customer churn increased in July," and the system translates these requests into database operations, statistical analyses, or trend evaluations.

The impact of this shift cannot be overstated. It allows decision-makers to explore information in real-time without relying on analysts or navigating complex interfaces. Analysts, in turn, benefit from a faster way of generating exploratory insights. The system handles the mechanical work of querying while users focus on interpretation. This creates a more natural, iterative analytical workflow in which questions lead to more questions, and insight emerges through dialogue rather than static reporting.

Conversational querying also reveals limitations. Ambiguous questions, incomplete context, and linguistic subtleties can lead to misinterpretation. To address this, advanced systems incorporate clarifying prompts, contextual learning, and grounding mechanisms that connect language to specific datasets, definitions, and organizational semantics. Effective natural language analytics therefore requires not only linguistic fluency but also structural alignment with the enterprise's data ecosystem.

11.4 Narrative Explanation and the Humanization of Insight

Beyond querying, natural language analytics excels in narrative explanation. The ability to convert numerical patterns into coherent stories represents one of the greatest strengths of generative AI. Instead of presenting users with charts alone, natural language systems contextualize the findings. They explain trends, highlight anomalies, articulate potential causes, and suggest next steps. This narrative capability transforms raw analytics into comprehensible insight.

Narrative explanation supports better decision-making because humans process stories more effectively than tables or graphs. When a model explains that "revenue declined due to reduced retention among mid-tier customers, driven by slower response times and increased competition," it integrates quantitative analysis with interpretive reasoning. Users receive not only numbers but meaning.

Narrative analytics also enhances accessibility. Non-technical employees can understand and interpret insights without deep statistical or technical training. Executives can absorb complex analyses quickly. Operational teams can make informed adjustments without deciphering dashboards. The system becomes an interpreter that bridges the gap between technical output and human understanding.

11.5 Multimodal Conversations with Data

Natural language analytics expands further when integrated with multimodal intelligence. A user can show the system a chart and ask what it means. The system can examine the trend, detect outliers, and provide a written explanation. A user can upload an image of a dashboard, and the system can interpret the metrics. A user can share a financial report and request a summary of key drivers.

In such interactions, language becomes the connective tissue that unites text, images, charts, documents, and data tables. The user no longer needs to switch between tools or formats. Insight emerges through a seamless conversation that blends modalities. This multimodal capability brings analytics closer to the way humans naturally work, where understanding arises from integrating multiple sources of information rather than analyzing each in isolation.

The result is a more natural analytical environment, one where the user interacts with data through observation, questioning, interpretation, and refinement, which is much like a conversation with a knowledgeable colleague.

11.6 Natural Language as an Enabler of Insight Democratization

Natural language analytics has the power to democratize insight by allowing individuals across the organization to interact with data in plain language rather than through specialized tools or technical interfaces. Employees who previously relied on analysts or dashboards can now ask questions directly, explore patterns, and receive explanations tailored to their level of expertise. This expanded accessibility helps organizations make decisions faster, involve more stakeholders, and cultivate a more data-informed culture.

Democratization, however, requires intentional design. When more people engage directly with data, systems must ensure that responses are grounded in the correct sources, that definitions remain consistent across teams, and that users understand the limitations of model-generated insight. Natural language systems reduce barriers to participation but do not eliminate the need for clarity, context, and shared understanding. When implemented thoughtfully, natural language analytics empowers the organization broadly while preserving coherence and reliability in how insights are interpreted and used.

11.7 Reliability Challenges and the Need for Grounding

Natural language analytics inherits the strengths and weaknesses of generative AI. While these systems can interpret questions, generate explanations, and translate analytical results into coherent narratives, they can also misinterpret queries, overlook contextual nuances, or produce confident statements that are not supported by underlying data. These reliability challenges become more pronounced as conversational interfaces reach non-technical users who may place unwarranted trust in fluent responses.

To ensure dependable insight, modern natural language systems incorporate grounding mechanisms that anchor responses directly to verified datasets, metadata, definitions, and organizational semantics. Rather than generating answers solely from learned patterns, grounded systems retrieve factual values, reference authoritative sources, and provide traceable links to the data that informed each response. Provenance tracking further allows users to verify inputs and validate conclusions, strengthening transparency and trust.

Even with grounding, human oversight remains essential. Natural language analytics is most effective when paired with critical thinking, contextual judgment, and domain expertise. Reliability emerges not from perfect automation but from a workflow in which humans and AI collaborate, with each compensating for the other's limitations while amplifying each other's strengths.

11.8 The Evolving Role of Analysts in a Conversational World

As natural language interfaces permeate organizations, the role of analysts evolves. Instead of focusing on manual querying or dashboard creation, analysts shift toward curating datasets, validating insights, designing analytical frameworks, and interpreting complex or ambiguous results. They become stewards of meaning rather than producers of static reports.

Natural language systems amplify analytical capacity rather than diminishing it. Analysts gain a powerful tool that handles repetitive tasks, extracts patterns quickly, and provides initial interpretations. This allows them to devote more time to strategic analysis, cross-functional communication, and the resolution of deeper analytical questions.

The interplay between human expertise and conversational intelligence becomes the foundation of modern analytics. Analysts guide, correct, and refine the AI's output, ensuring that insights are grounded and aligned with organizational context.

11.9 Conclusion: Toward Seamless, Conversational Intelligence

Natural language analytics signals a new era where insight flows through dialogue rather than dashboards. By integrating linguistic understanding, generative reasoning, and multimodal interpretation, these systems create an analytical environment that mirrors natural human communication. They empower individuals at every level of the organization, accelerate decision cycles, and transform analytics from a technical discipline into a conversational partnership.

As conversational interfaces become embedded into everyday workflows, analytics evolves from a specialized activity into an accessible, collaborative process, in which humans and intelligent systems reason together, each strengthening the other's capabilities.

CHAPTER 12

Data Preparation and Feature Engineering in Intelligent Systems

12.1 Introduction: The Hidden Architecture of Intelligent Models

Data preparation and feature engineering form the quiet foundation upon which all AI systems are built. While models and neural architectures draw the most attention, no system, regardless of sophistication, can overcome flawed or poorly understood data. AI does not learn directly from reality; it learns from the data that represents it. If that data is incomplete, inconsistent, biased, or poorly structured, the resulting system will inherit and amplify those flaws. Therefore, preparation is not a mechanical preprocessing step. Rather, it is the act of shaping how the AI system perceives the world.

Historically, data preparation was often underestimated and treated as routine technical work. The rise of machine learning, and especially deep learning, revealed how essential early-stage preparation truly is. In AI environments, data preparation determines not only performance but fairness, stability, and long-term reliability. Feature engineering, in turn, transforms raw information into meaningful signals that reflect the structure of the underlying problem. Even in deep learning, where models learn representations automatically, the quality and structure of inputs remain indispensable.

This chapter explores data preparation and feature engineering as foundational pillars of intelligent systems, including how they shape model performance, fairness, reliability, and long-term AI behavior. It also explains why preparation decisions often have greater impact than model selection itself.

12.2 Understanding the Challenge of Real-World Data

Real-world data rarely arrives in clean or coherent form. It is collected under varying conditions, shaped by legacy systems, influenced by human behavior, and subject to operational noise. Organizations often assume the difficulty of AI lies in developing complex models, but the most significant challenges typically emerge long before modeling begins. Raw data may include missing values, duplicated records, misaligned timestamps, inconsistent definitions, unbalanced outcomes, or patterns that reflect historical inequities.

These imperfections, which are not anomalies, are inherent to systems built over time by different teams, technologies, and priorities. AI systems trained on such data risk learning distorted representations of reality, by mistaking proxy variables for causation, failing to recognize minority patterns, or interpreting noise as meaningful structure. Understanding real-world data therefore requires careful profiling, contextual awareness, and collaboration between domain experts and data specialists. This combined expertise allows teams to separate meaningful signals from misleading artifacts and to detect biases that must be addressed early rather than after deployment.

12.3 Cleaning, Repairing, and Standardizing Data

Cleaning and repairing data form the essential groundwork upon which all intelligent systems depend. This stage involves resolving missing values, correcting structural inconsistencies, harmonizing formats, and identifying anomalies that may arise from data-entry errors, faulty sensors, or flawed system integrations. Although often viewed as routine, these tasks require judgment and contextual knowledge. A missing value may signal a simple omission or reveal deeper behavioral patterns such as customer disengagement, equipment malfunction, or operational delays. Likewise, an extreme value might represent a genuine rare event or a systemic error that should not be learned by the model.

Standardization is equally critical. Inconsistent definitions can be more damaging than incomplete records because inconsistent semantics introduce patterns that models internalize as truth. When the same metric is defined differently across business units, or redefined over time without documentation, models absorb these contradictions and produce unstable or misleading predictions. Rigorous preparation therefore requires alignment of definitions, units, formatting conventions, and underlying assumptions. This work not only enhances data reliability but also strengthens organizational trust in downstream insights, ensuring that model behavior rests on a coherent and stable foundation rather than shifting or ambiguous inputs.

12.4 Integrating Data Across Systems and Contexts

Modern organizations operate across numerous systems such as transactional databases, CRM platforms, ERP suites, IoT sensors, operational logs, and external data feeds, where each captures information through its own lens. However, intelligent systems require integrated views that reflect complete, context-rich narratives. Integration involves resolving identities across systems, aligning timeframes, reconciling conflicting attributes, and merging datasets with fundamentally different structures or collection methods. When executed well, integration enriches the information available to models and supports a more complete understanding of customers, operations, assets, and environmental conditions.

Integration is not solely a technical workflow; it is fundamentally conceptual. Customer-behavior data must be interpreted alongside operational constraints, product attributes, supply-chain events, or external factors. Without thoughtful integration, machine learning models may misinterpret interactions between systems as meaningful patterns, producing correlations that reflect dataflow artifacts rather than genuine relationships. Effective integration therefore requires careful deduplication, contextual alignment, harmonization of definitions, and thorough lineage tracking. When these practices are in place, fragmented data sources become unified intelligence, laying the conceptual and technical groundwork upon which accurate, adaptive, and trustworthy intelligent systems are built.

12.5 The Art and Purpose of Feature Engineering

Feature engineering is the process of transforming raw data into representations that highlight the underlying structure of a problem. Before deep learning, it was often the most creative and intellectually demanding aspect of building models, requiring analysts to encode domain knowledge into numerical form. This included constructing ratios, aggregating historical behavior, extracting trends, and designing indicators that helped models learn effectively. Although deep learning automates many representational tasks, feature engineering remains essential, especially in structured enterprise data where domain context plays a central role in shaping meaningful inputs.

A well-engineered feature distills complex phenomena into a form that models can use reliably. Raw timestamps become cyclical indicators that capture weekly or seasonal patterns. Sequences of interactions become behavioral summaries of churn or engagement. Sensor readings become predictive signals of wear or failure. Feature engineering bridges the gap between human understanding of a domain and the mathematical representations required for learning. Its effectiveness depends on domain insight, analytical creativity, and interdisciplinary collaboration, not on technical transformation alone.

12.6 Feature Engineering in the Deep Learning Era

Deep learning introduced automatic feature extraction for domains such as vision, speech, and natural language. Convolutional networks learn visual patterns without predefined shapes, and transformer models learn linguistic or temporal patterns without manual encoding. These architectures reduce the need for handcrafted features in unstructured data. However, automatic representation does not eliminate the need for careful preparation. Deep models still depend on well-structured, well-curated inputs, and they must receive data that is complete, representative, ethically sourced, and free from systematic bias.

In structured business data, feature engineering remains central. Deep learning models applied to tabular data often benefit from domain-informed transformations, historical aggregations, meaningful ratios, engineered indicators, and thoughtful encoding of categorical variables. While normalization is critical for numerical inputs in tabular or sensor contexts, text and other tokenized inputs require different preparation, which can include tokenization, vocabulary management, and appropriate sequence handling rather than numerical scaling. Thus, the deep learning era shifts the focus from manual feature creation toward the design of feature spaces, input structures, and representational strategies that allow models to learn effectively. Automatic feature learning reduces effort but does not replace the judgment required to prepare high-quality inputs.

12.7 Creating Temporal, Behavioral, and Contextual Features

Many of the most effective intelligent systems rely on features that represent time, behavior, and context. Temporal features capture how values evolve, such as how they rise, decline, accelerate, stabilize, or cycle over time. Behavioral features summarize sequences of actions, revealing habits, preferences, intent signals, and anomalies that static snapshots cannot capture. Contextual features situate events within broader conditions, such as seasonality, geography, operational workload, environmental factors, or economic climate. Together, these feature types allow models to understand dynamics rather than isolated observations.

Constructing such features requires careful design. Time-series models depend on consistent timestamp alignment, proper handling of irregular intervals, and suitable windowing techniques. Behavioral features require thoughtful aggregation, sequence encoding, or pattern extraction that preserves meaning without overwhelming models with noise. Contextual features often demand external data enrichment and precise alignment with internal timelines, events, and metrics to avoid temporal leakage or misinterpretation. The quality of these temporal, behavioral, and contextual abstractions influences whether a model captures deep structural relationships or remains confined to surface-level correlations. Well-crafted features bring nuance, realism, and explanatory depth to intelligent systems.

12.8 Detecting and Mitigating Bias During Preparation

Data preparation is the earliest and most critical opportunity to identify and mitigate bias. Bias typically emerges long before modeling, in variable definitions, data-collection practices, sampling processes, historical imbalances, and the omission of key populations. If these issues are overlooked, intelligent systems will reproduce and potentially amplify inequities present in the underlying data. Effective mitigation requires examining distributions across demographic groups, assessing representativeness, analyzing patterns of exclusion, and identifying whether certain features serve as indirect proxies for sensitive attributes.

Preparation is also the stage where decisions about data inclusion, exclusion, and transformation carry long-term ethical implications. Removing certain cases may simplify pipelines but distort the system's understanding of reality. Retaining variables that encode socioeconomic or demographic patterns may inadvertently introduce discriminatory behavior. Responsible preparation therefore requires transparency in decision-making, thorough documentation, and collaboration with domain experts, legal teams, and ethics groups. Addressing fairness considerations during preparation ensures that equitable outcomes are embedded in the foundation of the system rather than treated as corrective adjustments after modeling.

12.9 The Role of Documentation, Lineage, and Transparency

High-quality data preparation and feature engineering require rigorous documentation. Every transformation, such as a cleaning operation, aggregation, encoding, imputation, or feature creation, contributes to the lineage of the dataset. This lineage determines the model's ability to be audited, reproduced, understood, and trusted. In intelligent systems, documentation also becomes a core component of governance and compliance, especially in regulated industries where explainability, oversight, and traceability are mandatory.

Transparency in preparation ensures that downstream teams understand why a feature behaves in certain ways, how a dataset was constructed, and where potential vulnerabilities may lie. When documentation is absent or incomplete, intelligent systems become opaque, and failures become difficult to diagnose or correct. Proper lineage tracking transforms preparation from a hidden, ad-hoc process into a structured, inspectable pipeline that supports long-term reliability and organizational trust.

12.10 Conclusion: Preparation as the Foundation of Trustworthy Intelligence

Data preparation and feature engineering form the backbone of intelligent systems. They transform messy, imperfect, real-world information into coherent representations from which models learn. They determine the quality, fairness, reliability, and resilience of AI systems long before model training begins. Although model architectures evolve rapidly, the importance of preparation remains constant. Intelligent systems succeed when they are grounded in well-structured, well-understood, and ethically stewarded data.

CHAPTER 13

Model Development: Crafting Intelligent Behavior

13.1 Introduction: From Data to Intelligence

Model development is the phase in which data, engineered features, domain expertise, and algorithmic methodology merge to create functional intelligent systems. It is the point where abstract questions become computational formulations capable of producing predictions, recommendations, or decisions. While modern discussions often focus on sophisticated algorithms or advanced model architectures, the true intellectual foundation of model development extends far beyond technical complexity.

Effective model development involves translating organizational problems into formal structures, extracting patterns from curated data with discipline, and designing systems that can operate reliably in dynamic, real-world environments. As a result, effective model development demands technical proficiency, conceptual clarity, ethical awareness, and thoughtful attention to the broader sociotechnical context in which the model will function.

This chapter explores how models are designed, trained, tuned, and shaped into purposeful intelligent behavior. It examines how technical choices during development directly influence performance, reliability, fairness, and long-term system outcomes. It also clarifies how development practices encode both intended objectives and unintended risks.

13.2 Framing the Objective: Defining What Success Means

A model's utility is determined first and foremost by the clarity of its objective. Without a precise definition of the decision the model is intended to support, development efforts risk producing systems that are technically sound yet operationally irrelevant. Defining the objective requires an in-depth understanding of the organizational context, the stakeholders involved, and the practical implications of model-driven decisions. Success must be articulated not in strictly quantitative terms but in relation to the strategic purpose the model serves.

For example, a customer retention model is not merely expected to predict churn; it must do so with sufficient lead time to enable intervention. A medical triage system must reflect clinically meaningful prioritization rather than maximizing predictive accuracy alone. A predictive maintenance model must identify failures early enough to prevent operational disruption. Objective framing also necessitates explicit consideration of ethical implications. A narrowly optimized metric may yield impressive performance while generating unintended social or operational harm. Responsible model development begins with the establishment of an objective that is precise, contextualized, and ethically bounded.

13.3 Selecting the Appropriate Modeling Approach

Once a problem has been articulated with clarity, the appropriate modeling approach must be selected. This decision depends on the nature of the data, the complexity of the underlying relationships, and the operational and regulatory constraints of the domain. Traditional statistical models may provide transparency and stability in contexts requiring interpretability. Machine learning models may capture nonlinear relationships in structured data, while deep learning or foundation-model-based approaches may be necessary for unstructured or multimodal inputs.

Model selection is inherently iterative. Initial experiments often reveal limitations or practical constraints that necessitate revisiting earlier assumptions. A model demonstrating exceptional performance under controlled conditions may nevertheless prove unsuitable for production if it depends on fragile features, exhibits instability under minor data shifts, or requires inputs unavailable at inference time. Thus, selecting a modeling approach is simultaneously an empirical process and an exercise in practical judgment, balancing predictive capability, interpretability, operational feasibility, and ethical constraints.

13.4 Training: Teaching Systems to Recognize Patterns

Training is the computational process through which a model learns patterns from data. In this phase, the model generates predictions, compares them to observed outcomes, and adjusts its internal parameters through iterative optimization. Despite the automated nature of this process, training is guided by a series of design decisions that fundamentally influence model behavior. These include determining how data should be partitioned, how class imbalances should be managed, how regularization should be applied to prevent memorization, and how potentially sensitive attributes should be controlled to prevent discriminatory outcomes.

Training frequently reveals limitations in the underlying dataset. Persistent difficulties in convergence may indicate missing features, inconsistent labeling practices, unrepresentative samples, or conceptual misalignment between the objective and the available data. In this sense, training is both an act of model construction and a diagnostic tool that deepens understanding of the data's underlying structure and the assumptions embedded within it.

13.5 Validation: Ensuring the Model Actually Understands

Validation assesses a model's capacity to generalize to previously unseen data. This distinction between learning and memorization is central to determining whether a model demonstrates genuine inferential capability. A model that performs well on training data but poorly on validation data is said to be overfitting, capturing superficial patterns that do not reflect the broader domain.

Effective validation extends beyond aggregate accuracy. It evaluates behavior across demographic groups, temporal segments, and operational scenarios to ensure that performance is stable and equitable. Validation may expose vulnerabilities requiring revision of features, reconsideration of the objective, or restructuring of the data. The iterative nature of validation strengthens the model's robustness and ensures that it behaves reliably in the heterogeneous and evolving environments characteristic of real-world deployment.

13.6 Navigating the Trade-Off Between Accuracy and Interpretability

As model complexity increases, interpretability often declines. Deep neural networks and ensemble methods may provide superior predictive performance yet behave as opaque systems whose internal logic is not readily comprehensible. While opacity may be tolerable in certain domains, in high-stakes contexts such as healthcare, credit evaluation, or legal judgment, lack of interpretability can undermine trust, compliance, and ethical accountability.

Model development must therefore incorporate interpretability as a design requirement rather than an optional enhancement. In some cases, simpler but more transparent models may be preferable. In others, post-hoc explanation techniques may provide approximate insight into the reasoning of complex models, though such explanations must be treated with caution. The choice reflects not only technical considerations but organizational values, legal requirements, and broader societal expectations. It is a normative judgment regarding the balance between predictive performance and responsible transparency.

13.7 Addressing Bias, Fairness, and Ethical Risks

Models inherit patterns present in training data and, therefore, they inevitably reflect the social and historical dynamics embedded within those data sources. If the data disproportionately represents certain groups, encodes structural inequities, or contains proxy variables correlated with sensitive attributes, the model may generate unfair or harmful outcomes.

Addressing these issues requires systematic examination of model behavior across demographic subgroups, rigorous assessment of the influence of specific features, and explicit evaluation of whether the optimized objective aligns with ethical expectations. Fairness interventions may require adjusting sampling procedures, modifying feature sets, redesigning objectives, or incorporating human review. In some cases, the most responsible course of action may be to reject automation entirely. Ethical considerations thus permeate every stage of model development, linking technical decisions to broader organizational and societal consequences.

13.8 Stress Testing and Scenario Analysis

Models deployed in real-world environments must be resilient to variability that exceeds the conditions observed during training. Stress testing exposes the model to extreme values, atypical patterns, and hypothetical disruptions to evaluate its stability under adverse conditions. Scenario analysis extends this process by assessing how changes in underlying assumptions or external environments impact model performance.

These practices transform model development from a static exercise into a dynamic assessment of adaptive capability. A model that performs reliably under stress is more likely to maintain stability in the unpredictable and evolving conditions characteristic of production environments.

13.9 Human Expertise as an Essential Component

Model development is inherently interdisciplinary. Data scientists contribute algorithmic expertise, domain experts provide contextual depth, engineers ensure operational viability, and ethics and legal teams safeguard compliance and responsible use. Decision-makers, meanwhile, determine whether the model meaningfully advances organizational objectives.

This collaboration ensures that the resulting system is mathematically rigorous, contextually grounded, ethically responsible, and operationally sustainable. Human judgment remains indispensable, guiding interpretation, overseeing risks, and validating whether the model's behavior aligns with the organization's mission and values.

13.10 Conclusion: Crafting Models That Behave Intelligently and Responsibly

Model development is the disciplined construction of intelligent behavior. It transforms curated data into functional predictions and complex signals into actionable insight. Yet the value of a model is not defined solely by its accuracy. It rests equally upon reliability, fairness, interpretability, resilience, and alignment with broader organizational and societal expectations.

As the work of development transitions to the subsequent phase of rigorous evaluation and validation, a central lesson that persists is that intelligent systems are not merely engineered. They are deliberately crafted. Their behavior reflects the quality of the data, the clarity of the objectives, the discipline of the development process, and the integrity of the human judgment guiding their creation.

CHAPTER 14

Evaluation, Validation, and Benchmarking AI Models

14.1 Introduction: Ensuring Models Are Ready for the Real World

Evaluation is the stage at which a model's performance, limitations, and internal assumptions are made visible. Whereas development focuses on learning patterns from historical data, evaluation tests whether those learned patterns generalize to novel, unseen conditions. This distinction separates genuine inferential capability from superficial memorization. Evaluation serves as the safeguard that ensures models behave predictably, ethically, and reliably before they are entrusted with real-world decisions.

While training establishes the potential for intelligent behavior, evaluation determines whether that behavior can be trusted outside controlled environments. A system that performs exceptionally well during development may exhibit critical weaknesses when exposed to operational variability. For this reason, evaluation is not a single event at the end of the pipeline but a continuous methodological discipline embedded throughout the model lifecycle. It considers statistical performance, robustness, fairness, temporal stability, and alignment with organizational purpose.

This chapter explores the multifaceted nature of AI model evaluation, including how validation, benchmarking, fairness assessment, and robustness testing ensure reliability, trust, and responsible deployment.

It also explains how evaluation acts as the connective tissue between technical performance and real-world impact.

14.2 The Purpose of Evaluation: Beyond Numerical Accuracy

Although evaluation often begins with quantitative metrics such as accuracy, error rates, or loss functions, meaningful validation extends far beyond these numerical indicators. Metrics may summarize performance, but they do not determine whether a model is suitable, actionable, or safe for its intended context. A system may achieve high accuracy yet behave in ways that are brittle, biased, or misaligned with organizational objectives.

The broader purpose of evaluation is to assess whether a model's outputs contribute to improved decisions. This assessment requires a deep understanding of the operational environment. A predictive maintenance model must be evaluated not only for its precision but for its impact on production continuity. A clinical triage model must be judged not only by statistical measures but also by its safety implications and ethical acceptability. A customer churn model must demonstrate actionability, not simply numerical accuracy. Evaluation therefore integrates technical performance with organizational relevance, situational awareness, and ethical stewardship.

14.3 Distinguishing Training, Validation, and Testing

The division of data into training, validation, and testing subsets is a foundational principle of model development, yet its significance is often underestimated. The training set provides the basis for learning. The validation set supports model refinement, hyperparameter tuning, and selection among candidate approaches. The test set, held entirely separate from both training and validation, provides an unbiased assessment of performance on previously unseen examples.

Maintaining clear separation among these datasets is essential to ensuring the integrity of evaluation. When data leaks across boundaries, through duplication, improper partitioning, or repeated tuning cycles, a model may appear more capable than it actually is. Contamination erodes the credibility of evaluation and can mask profound weaknesses. A rigorous separation among training, validation, and test data ensures that performance reflects actual generalization rather than unintentional memorization or artifacts of the modeling process.

14.4 Understanding Error: What Models Get Wrong and Why

Analyzing errors is often more instructive than examining successes. Errors reveal systematic weaknesses, blind spots, and inconsistencies in how the model interprets data. Persistent mispredictions for particular subgroups, conditions, or contexts may indicate gaps in representation, flaws in feature design, or distortions arising from historical patterns embedded in the data.

Error analysis requires distinguishing between random and systematic error. Random errors are often acceptable in domains characterized by inherent uncertainty. Systematic errors, by contrast, highlight deeper structural issues. If a fraud detection system disproportionately flags legitimate transactions from certain regions, or if a diagnostic model consistently underestimates severity for particular age groups, these patterns reflect problems that must be addressed. Through careful and contextual error analysis, developers gain insight into where models require refinement, where data needs augmentation, and where human oversight must remain integral.

14.5 Fairness Evaluation: Ensuring Equitable Outcomes

Fairness evaluation is a critical dimension of assessing intelligent systems. Models frequently influence decisions that impact individuals and communities, and uneven performance across demographic groups can produce harmful inequities. Fairness evaluation examines whether particular populations experience disproportionate error rates, whether sensitive attributes influence predictions directly or indirectly, and whether historical biases are being reproduced or amplified.

Fairness is inherently multidimensional and cannot be reduced to a single metric. It involves legal, ethical, cultural, and societal considerations and requires interdisciplinary assessment involving domain experts, legal advisors, community stakeholders, and ethicists. A fair model is not one that performs identically across all groups but one that aligns with organizational values, adheres to regulatory requirements, and behaves in ways that are socially responsible. Fairness evaluation ensures that AI systems operate equitably and do not perpetuate or exacerbate existing disparities.

14.6 Stress Testing, Robustness, and Sensitivity Analysis

Real-world environments are dynamic and often volatile. Stress testing exposes models to extreme or atypical inputs to assess their resilience to unexpected conditions. This process is essential in fields where rare events carry significant impact, such as financial risk assessment, cybersecurity, infrastructure monitoring, or public health. A model that performs well under average conditions but collapses under stress is unsuitable for deployment.

Robustness evaluation examines a model's stability under plausible variations. Sensitivity analysis complements robustness testing by evaluating how small perturbations in input values influence outputs. A reliable system should not respond erratically to minor fluctuations. Sensitivity analysis reveals dependencies on unstable or noisy features, highlighting where the model may require simplification, feature redesign, or enhanced regularization. Together, stress testing and sensitivity analysis ensure that the model behaves predictably across both expected and unforeseen scenarios.

14.7 Evaluating Temporal Stability and Drift

Data distributions evolve over time, and a model that performs effectively today may degrade as patterns shift. Evaluating temporal stability involves examining model performance across historical and prospective time windows to determine whether the system maintains validity as conditions change. Concept drift may arise from changes in customer behavior, market dynamics, environmental conditions, operational processes, or broader socioeconomic factors.

Evaluation must therefore assess not only static performance but longitudinal reliability. Degradation over time may indicate the need for retraining, recalibration, or even reconceptualization of the model. Understanding and monitoring drift is crucial for maintaining reliability in dynamic environments and ensuring that intelligent systems remain adaptive rather than becoming obsolete or harmful.

14.8 Benchmarking: Comparing Models to Alternatives

Benchmarking situates a model's performance within a broader landscape of possible approaches. By comparing a candidate model to simpler baselines, prior versions, alternative algorithms, or industry standards, benchmarking helps determine whether the added complexity of a new system yields meaningful benefit. Without benchmarking, organizations risk adopting models that increase operational burden without improving outcomes.

Benchmarking clarifies trade-offs among accuracy, interpretability, computational efficiency, fairness, and maintainability. A simpler model may provide nearly identical performance while offering greater transparency and ease of audit. A more complex model may justify its adoption only if it demonstrates substantial and contextually meaningful gains. Benchmarking ensures that model selection reflects a deliberate evaluation of alternatives rather than an unexamined preference for novelty or complexity.

14.9 When Evaluation Reveals That a Model Should Not Be Deployed

Responsible AI practice requires recognizing when a model is not suitable for deployment. Evaluation may identify unacceptable biases, insufficient robustness, poor temporal stability, or misalignment with organizational goals. In some cases, the underlying problem may be inappropriate for automation; in others, the data may be too limited or too biased to support responsible modeling.

Rejecting a model at the evaluation stage is a sign of process integrity rather than failure. It prevents downstream operational risks, reputational harm, regulatory liability, and potential harm to individuals or communities. Evaluation ultimately functions as a checkpoint that protects against premature deployment and ensures that only systems meeting rigorous standards proceed to production.

14.10 Conclusion: Evaluation as the Gatekeeper of Trust

A rigorous evaluation process is indispensable for ensuring that intelligent systems behave as intended. It verifies that models are not only statistically sound but reliable under changing conditions, robust to uncertainty, and equitable in their impact. Evaluation functions as the final safeguard, which affirms that a model reflects organizational values, adheres to ethical principles, and can be trusted to operate responsibly in real-world environments. Only when a model meets these standards of reliability, fairness, and stability can it be considered ready for deployment.

CHAPTER 15

Deploying AI: Operationalizing Intelligence at Scale

15.1 Introduction: The Moment Intelligence Meets Reality

Deployment marks the point at which an intelligent system leaves the controlled domain of development and enters the complex, unpredictable conditions of real organizational environments. It is here that models begin influencing decisions, interacting with users, shaping operations, and producing real-world consequences. While model development focuses on extracting patterns from historical data, deployment determines whether those patterns translate into stable, actionable, and reliable behavior at scale. Therefore, the transition from experimentation to operation is not merely technical, it is organizational, cultural, and strategic.

Many organizations underestimate the challenges of deployment. They assume that once a model is trained, operationalization is straightforward. In practice, deployment is among the most demanding stages of the AI lifecycle. It requires infrastructure capable of supporting real-time or large-scale inference, integration with existing enterprise systems, robust monitoring frameworks, version control, governance processes, and the design of feedback loops that allow models to evolve.

This chapter explores what it takes to operationalize AI responsibly, including how deployment, monitoring, governance, and organizational integration ensure that intelligence becomes a stable and trustworthy component of decision-making. It also clarifies how breakdowns at this stage often determine whether AI succeeds or fails in practice.

15.2 From Prototype to Production: Understanding the Transition

Most models begin as exploratory prototypes developed in isolated environments and tested on controlled datasets. These prototypes are valuable tools for research and experimentation, but they rarely reflect production realities. The data pipeline differs significantly; latency, uptime, and security requirements become more rigorous; environmental noise and unanticipated inputs challenge model robustness.

Transitioning to production requires reconsidering the assumptions made during development. Manual data cleaning performed during experimentation must be replaced with automated, repeatable transformations. Feature-engineering pipelines must be fully reproducible. Inputs must be validated to prevent corrupted data from generating erroneous predictions.

Production deployment thus demands the creation of an ecosystem, including data flows, orchestration mechanisms, monitoring tools, and governance structures, that support reliable operation. This ecosystem forms the foundation of MLOps, the discipline that unifies machine learning with software engineering and operational excellence.

15.3 The Role of MLOps: Sustaining Intelligence Over Time

MLOps provides the operational framework through which machine learning systems become reproducible, maintainable, and adaptable. Similar to how DevOps transformed software delivery, MLOps establishes practices for continuous integration, versioning, automated deployment, model monitoring, and life-cycle management. It acknowledges that models are dynamic artifacts and that their performance shifts as the world changes, necessitating oversight and continual refinement.

A core principle of MLOps is reproducibility. Every step in the development process, including data preparation, feature transformations, and hyperparameter settings, must be captured, versioned, and replicable in production. Even small inconsistencies between development and production environments can degrade model behavior. MLOps also establishes real-time monitoring to detect drift, performance decay, latency issues, and fairness concerns. When drift is detected, models must be retrained, recalibrated, or sunset. In this sense, MLOps sustains the long-term stability and relevance of intelligent systems.

15.4 Building the Deployment Pipeline: From Data to Decision

A production AI system relies on a deployment pipeline that ensures data flows reliably from source to model and from model outputs to the systems or individuals who rely on them. This pipeline must transform raw data into the required formats, apply feature transformations consistently, validate inputs, handle anomalies, and ensure that missing or corrupted elements do not compromise predictions.

Deployment pipelines must meet the operational demands of the environment. For example, fraud detection models may require millisecond-level response times, credit scoring models must operate with high availability, and predictive maintenance systems must process continuous sensor feeds. These constraints shape architecture decisions, including whether to employ batch processing, streaming platforms, serverless components, distributed infrastructure, or edge deployment.

Version control and safe deployment practices are essential. Organizations often rely on shadow deployments, canary releases, rollback mechanisms, and A/B testing to evaluate new models before full release. These methods ensure that updates improve performance without destabilizing production systems.

15.5 Integrating AI with Enterprise Applications and Workflows

An intelligent system provides value only when it is meaningfully integrated into the workflows, applications, and decision processes it is designed to support. A highly accurate model that is not embedded in operational systems, user interfaces, or decision logic may remain unused.

Effective integration requires understanding how decisions are made within the organization. Predictions must reach the appropriate individuals or systems at the right moment and in the appropriate form. Integration strategies range from APIs that expose model predictions to business applications, to embedding intelligence within dashboards, decision engines, workflow automation tools, or real-time IoT systems operating at the edge.

Successful integration requires close collaboration among engineers, analysts, domain experts, and operational teams to ensure that outputs are interpretable, actionable, and aligned with the decision contexts in which they will be used.

15.6 Managing Risk During Deployment

Deployment introduces operational, technical, ethical, and organizational risks that may not be apparent during development. Models may behave unexpectedly when exposed to new environments, upstream system changes, or adversarial conditions. Faulty or corrupted data inputs can propagate errors rapidly. Ethical risks, such as biased outcomes or lack of transparency, become more pronounced when models operate at scale.

Effective risk management requires formal governance frameworks that define approval processes, guardrails, monitoring requirements, and escalation procedures. High-stakes deployments, such as those in finance, healthcare, or public safety, must undergo more rigorous validation, including fairness assessments, stress tests, and ethical reviews. Therefore, deployment is a governance milestone as much as a technical achievement.

15.7 Ensuring Human Collaboration and Oversight

Even in highly automated environments, human judgment remains essential. Deployment must define when human review is required, how users interpret AI outputs, and where responsibility resides when decisions involve both human and machine contributions. Human-in-the-loop and human-on-the-loop designs ensure that the most consequential decisions receive appropriate oversight.

Clear communication and training are critical. Users must understand the model's purpose, limitations, and appropriate use. Without this, they may place unwarranted trust in automated outputs or disregard them altogether. Deployment therefore includes developing documentation, conducting training, and establishing channels for feedback so that humans remain informed and engaged collaborators.

15.8 Monitoring the Model After Deployment

Deployment marks the beginning of a new operational phase in which the model's performance must be continuously monitored. External conditions evolve, data distributions shift, and model behavior can degrade over time. Continuous monitoring evaluates not only predictive performance but also fairness, stability, latency, coverage, and error patterns.

When drift or degradation is detected, organizations must initiate retraining, recalibration, or redesign. This process requires fresh data, validated pipelines, and controlled experimentation. Therefore, monitoring creates the feedback loop that connects operational environments back to development, enabling the model to remain reliable, safe, and effective throughout its lifecycle.

15.9 Conclusion: Deployment as a Living Commitment

Deploying an intelligent system is not a singular event but an enduring commitment to responsible operation. It demands technical rigor, organizational preparedness, governance maturity, and continuous oversight. Deployment transforms models from experimental artifacts into active components of enterprise decision-making, with real impacts on people, processes, and outcomes. For AI to remain a trustworthy part of the organization, deployment must be approached as an ongoing practice rooted in reliability, transparency, and ethical responsibility.

CHAPTER 16

Monitoring, Drift Detection, and Continuous Learning

16.1 Introduction: When Models Leave the Lab and Face the World

Once deployed, an intelligent system enters an environment defined by continual change. Customer preferences evolve, markets fluctuate, operational processes shift, and external forces, such as regulatory, social, and technological, reshape conditions in ways that cannot be fully anticipated during development. A model that performs with apparent perfection in a controlled setting may begin to degrade as soon as it encounters novel patterns, unobserved contexts, or subtle variations in incoming data. For this reason, monitoring, drift detection, and continuous learning constitute the foundation of responsible AI operations. They ensure that an intelligent system remains reliable not only at the moment of deployment but throughout its operational life.

Organizations often misinterpret deployment as a conclusion rather than a commencement. In reality, it marks the beginning of the most demanding phase of the AI lifecycle, which is ongoing performance management. Models are not static artifacts. They adapt, age, and interact with their environment in ways that require structured oversight.

This chapter examines the methods through which organizations preserve model integrity over time, emphasizing vigilant monitoring, early detection of degradation, and disciplined approaches to continuous adaptation. It also explains how continuous learning enables systems to remain accurate and relevant in changing environments.

16.2 The Central Role of Monitoring in Production AI

Monitoring is the systematic observation of a model's behavior under real-world conditions to determine whether it continues to function as intended. Unlike traditional software systems, which behave deterministically, intelligent systems are probabilistic. Their outputs depend as much on the quality and distribution of incoming data as on their internal structure. Effective monitoring must therefore extend far beyond technical uptime or system health metrics. It must assess whether predictions remain meaningful, stable, and aligned with the organization's strategic objectives.

Model monitoring encompasses multiple dimensions, including predictive accuracy, output stability, latency, data quality, fairness, and user interaction patterns. These indicators provide early visibility into emerging issues and help teams determine whether the model still reflects current reality. Because environments evolve, monitoring cannot remain static and it must adapt to new conditions and surface deviations promptly. The durability of an intelligent system ultimately depends on the strength of its monitoring framework.

16.3 Data Drift: When the World Changes Faster Than the Model

Data drift arises when the distribution of new incoming data diverges from the data used during training. Sometimes this shift is gradual, driven by seasonal patterns, demographic changes, or evolving user behavior. At other times it is abrupt, triggered by economic events, policy adjustments, interface redesigns, or new data collection methods. Even subtle deviations can accumulate and erode model reliability, often without immediate visibility. As a result, models in domains such as fraud detection, demand forecasting, or healthcare may silently degrade as their environments evolve.

Detecting and managing drift requires comparing current input distributions to historical baselines, assessing deviations that exceed normal variability, and interpreting their operational impact. Not all drift is harmful, some simply reflects natural change, but distinguishing benign variation from meaningful shifts is essential to maintaining model performance. Effective drift detection ensures that organizations can adapt their models before errors escalate into significant operational or financial consequences.

16.4 Concept Drift: When Relationships Themselves Evolve

Concept drift occurs when the fundamental relationships between inputs and outcomes evolve, even if the data distributions appear unchanged. It signals deeper shifts in how features relate to behaviors or decisions. Examples include customer churn dynamics altering with new competitors, economic uncertainty reshaping credit-risk relationships, or medical diagnostic patterns shifting as treatments or conditions evolve. In these cases, the model's original structural assumptions no longer hold, leading to declining performance despite seemingly stable inputs.

Addressing concept drift requires going beyond statistical comparisons to continuously evaluate predictive accuracy, subgroup stability, and consistency with domain expertise. Because concept drift alters the meaning of relationships rather than the data itself, it often prompts teams to re-examine not just the model but the underlying problem formulation. Effective monitoring ensures that models remain aligned with real-world dynamics rather than outdated assumptions.

16.5 Monitoring Model Performance in Dynamic Environments

Monitoring predictive performance requires access to ground truth, the actual outcomes against which predictions are evaluated. In many real-world scenarios, however, ground truth may not be available immediately. Fraud labels require investigation, medical diagnoses evolve over time, and predictive maintenance failures may occur infrequently. As a result, performance monitoring must operate on both short-term and long-term timescales.

Short-term monitoring examines proxies, decision consistency, anomaly patterns, and output stability. Long-term monitoring verifies predictive alignment as outcomes become available. Together, these perspectives reveal whether performance remains within acceptable thresholds or whether degradation is emerging. Monitoring must also examine subgroup-level behavior, as uneven degradation across populations may indicate emerging bias or structural drift.

16.6 Detecting Anomalies in Predictions and Behavior

Anomalies in predictions may signal early-stage failure. Sharp deviations in output distributions, unexplained clustering, abrupt changes in confidence scores, or inconsistent responses to similar inputs can indicate data corruption, upstream system failures, adversarial interference, or internal instability. Therefore, anomaly detection is an essential component of monitoring.

Interpreting anomalies requires contextual understanding. A sudden surge in predicted failures may reflect real deterioration or may reveal a malfunctioning sensor. A drop in approval scores may result from genuine shifts in applicant characteristics or from a data pipeline disruption. Domain expertise is indispensable for distinguishing between meaningful signals and artifacts.

16.7 Feedback Loops and the Importance of Non-Stationary Data

In many environments, a model's predictions influence the very data on which future predictions are based, a dynamic known as a feedback loop. When a model reduces supply in anticipation of low demand, future data will reflect the reduced supply rather than actual demand. When retention efforts focus on a subgroup identified as high-risk, future churn data may no longer reflect natural behavior.

Such feedback loops can create non-stationary environments, where the presence of the model alters underlying dynamics. Therefore, monitoring must account for the ways in which model-driven actions reshape data, potentially reinforcing narrow patterns or creating blind spots. Understanding feedback effects is essential for sustaining long-term model relevance.

16.8 Continuous Learning: When and How Models Should Adapt

Continuous learning involves updating models with new data to keep them aligned with changing conditions, but it is not universally appropriate. While frequent updates can increase responsiveness, they may also introduce instability, amplify noise, or embed assumptions that have not been properly vetted. Automatic retraining can create additional risks in regulated environments by undermining transparency, fairness, or auditability.

Implementing continuous learning effectively requires a disciplined framework that defines when retraining should occur, how new data should be selected and validated, and what safeguards ensure model updates are thoroughly tested before deployment. The goal is to balance adaptability with accountability, ensuring that models evolve in ways that enhance fidelity without undermining stability, governance, or ethical integrity.

16.9 Governance, Accountability, and Human Oversight in the Monitoring Process

Monitoring and drift detection require clear roles, responsibilities, and governance structures. Organizations must determine who monitors performance, who interprets drift signals, who authorizes retraining, and who intervenes when unexpected behavior emerges. Without defined accountability, monitoring becomes inconsistent and reactive.

Governance frameworks should document performance thresholds, escalation paths, retraining schedules, drift events, and communication processes. Transparency is essential for technical teams, as well as for users who must understand when and why model behavior changes. Human oversight integrates ethical judgment with technical evaluation, ensuring that monitoring reflects organizational values and societal expectations, not merely statistical metrics.

16.10 Conclusion: Sustaining Intelligence Through Vigilance

Monitoring, drift detection, and continuous learning are essential to sustaining the reliability and integrity of intelligent systems. These practices ensure that models evolve alongside the environments they serve, adapt to shifting conditions, and maintain fairness, safety, and stability over time. Intelligent systems are not autonomous entities, as they require continual stewardship. Therefore, monitoring is not a corrective procedure but a core commitment, one that underpins responsible, resilient, and trustworthy AI.

CHAPTER 17

AI System Risks: Fragility, Hallucination, and Emergent Behavior

17.1 Introduction: The Hidden Vulnerabilities of Intelligent Systems

As intelligent systems become embedded in the core operational and decision-making processes of organizations, the significance of their vulnerabilities grows proportionally. What was once a supplemental analytical tool has evolved into a decision-support layer that influences business operations, public policy, healthcare delivery, financial services, and social infrastructures. This expanded role increases not only the benefits of AI but also the scale and consequences of its failures. Errors that were once localized now propagate across interconnected systems, amplifying both technical and organizational risk.

Intelligent systems are not deterministic machines; they are statistical, adaptive, and context-dependent. These properties yield remarkable capabilities, but they also introduce forms of fragility that do not resemble the failure modes of traditional software. Small data shifts, subtle feedback loops, or unexpected interactions between models and environments can trigger disproportionate effects. As a result, understanding and managing these unique vulnerabilities becomes essential for building AI systems that remain reliable, resilient, and aligned with human expectations.

This chapter explores foundational AI system vulnerabilities, including fragility, hallucination, and emergent behavior, and examines how these risks arise from the structural nature of learning systems and affect reliability at scale. It also explains why these vulnerabilities intensify as AI systems become more autonomous and interconnected.

17.2 The Fragility of Statistical Intelligence

AI systems learn statistical patterns rather than true conceptual understanding, which makes them highly effective when real-world conditions mirror their training data but vulnerable when they encounter unfamiliar or anomalous situations. Fragility arises whenever inputs fall outside the learned manifold, scenarios for which the model has no precedent. As a result, fraud-detection models may falter when tactics change, predictive maintenance systems may misinterpret rare sensor anomalies, and recommendation engines may degrade as preferences shift unexpectedly.

These failures are not malfunctions but natural consequences of statistical approximation, where even small deviations from historical patterns can trigger significant performance drops, especially in high-stakes applications. Recognizing this fragility is essential for building monitoring systems, fallback strategies, and operational safeguards that ensure intelligent systems remain reliable in dynamic and uncertain environments.

17.3 Distribution Shift: When the World Deviates from the Past

Distribution shift, also known as covariate shift, occurs when the statistical properties of new data differ from those seen during training. While traditional software systems tolerate minor input variations without major consequences, machine learning models can behave unpredictably even under modest deviations. Such shifts may stem from evolving customer behavior, economic disruptions, sensor degradation, interface redesigns, regulatory changes, or modifications in data-collection processes.

Organizational changes, such as new product lines or updated workflows, can also introduce unexpected variations.

The danger of distribution shift lies in its subtle, often undetected progression. A model may continue producing outputs that seem plausible even as its internal representations grow increasingly misaligned with real-world conditions. Without systematic monitoring and temporal validation, performance can erode quietly until it results in operational failures or distorted decision-making. Recognizing that models trained on historical data remain continuously exposed to changing environments is essential for maintaining reliability over time.

17.4 Hallucination: When AI Generates Confident Misrepresentations

Generative AI introduces a unique risk through the creation of fluent but inaccurate outputs known as hallucinations. These occur when a model produces statements that are linguistically polished and delivered with confidence yet are factually wrong, logically inconsistent, or unsupported by evidence. The root cause lies in how generative models function. They predict the most probable sequence of tokens, not the sequence that is true. Their responses reflect statistical patterns in training data rather than a grounded understanding of external reality, leading them to fabricate plausible but incorrect information when asked questions beyond their knowledge or requiring factual verification.

This risk is intensified by users' natural tendency to equate fluency with correctness, giving hallucinated statements an undeserved aura of authority. Because of this, hallucinations can mislead users, distort decisions, or introduce false justifications into analytical workflows. Mitigating these failures requires grounding responses in verified data, using retrieval-augmented methods, applying output verification steps, and communicating uncertainty clearly.

17.5 Apparent Reasoning Versus Genuine Understanding

Modern AI systems often generate outputs that resemble structured reasoning, such as step-by-step explanations, causal analyses, or logical arguments. However, these should not be misconstrued as genuine reasoning. Such outputs are reconstructions of linguistic patterns learned during training rather than the result of internal conceptual models, causal understanding, or intentional inference. Because models lack a persistent world model and do not reason about causes or consequences, they frequently produce "post hoc rationalizations," which are explanations that sound convincing but bear no relationship to the actual computational process that generated the prediction.

This gap between appearance and reality creates significant risks, especially in domains requiring interpretability or auditability. Users may over-trust an AI system simply because its explanations seem coherent, leading to automation bias and uncritical acceptance of flawed outputs. Recognizing that these systems simulate reasoning rather than perform it is essential for maintaining responsible oversight, ensuring that human judgment remains central, and preventing misplaced confidence in explanations that only mimic understanding.

17.6 Emergent Behavior: Intelligence That Was Never Designed

As AI models increase in scale and complexity, they develop emergent behaviors, capabilities or failure modes that were never explicitly programmed but arise from interactions among billions of parameters. Some emergent behaviors are beneficial, such as spontaneous translation abilities or generalized pattern recognition, while others are unpredictable, brittle, or harmful. High-dimensional optimization enables models to discover internal representations that humans cannot fully interpret, making both desirable capabilities and unexpected shortcuts possible.

Because these systems operate in ways that exceed human foresight, emergent behavior highlights the difficulty of predicting model performance across diverse scenarios. Models may behave strategically, optimize in unintended ways, or exploit loopholes in objective functions, producing outputs that satisfy evaluation metrics while violating underlying intent. Addressing these risks requires continuous evaluation, adversarial testing, and institutional humility, acknowledging that complex AI systems can surprise even their creators and must be governed with ongoing scrutiny rather than static assumptions.

17.7 Automation Bias: When Humans Trust Machines Too Much

Even when AI models behave correctly, human perception can magnify risks. Automation bias occurs when users place excessive trust in machine-generated recommendations, often assuming greater precision, objectivity, or reliability than warranted. This bias is exacerbated by the persuasive fluency of modern systems and by interfaces that imply authority or finality.

In domains such as clinical decision support, hiring, lending, or public safety, automation bias can lead to uncritical acceptance of model outputs, suppression of expert judgment, or institutionalization of subtle model errors. Mitigation requires designing systems that communicate uncertainty, encourage human review, and clarify the appropriate role of AI within decision processes.

17.8 Cascading Failures in AI-Driven Environments

AI systems operate within interconnected technical and organizational ecosystems. Failures rarely remain isolated. A flawed forecast can propagate into supply chain disruptions, customer dissatisfaction, or distorted data that affects downstream models. Risk scores that misclassify individuals may reshape future datasets, reinforcing the initial misclassification in a feedback cycle.

Cascading failures arise because intelligent systems influence the environments from which they learn. When these systems interact across organizational layers, small errors can amplify into large-scale disruptions. Preventing cascading failures requires holistic system-level thinking, cross-model evaluation, and governance structures that recognize interdependencies across the AI ecosystem.

17.9 The Illusion of Stability in Complex AI Systems

Intelligent systems may appear stable for extended periods even as underlying vulnerabilities accumulate. This illusion arises because degradation often occurs gradually or in dimensions that are not directly observable. A model may behave consistently until a sudden shift exposes long-latent weaknesses. Updating adjacent systems, modifying user workflows, or introducing new data sources can trigger abrupt and destabilizing failures.

Sustaining reliability requires recognizing that stability must be actively maintained. Continuous monitoring, drift detection, periodic audits, adversarial evaluation, and scenario testing are necessary to prevent unpredictable failures and ensure that complex systems remain aligned with their intended purpose.

17.10 Conclusion: Managing the Behavior of Intelligent Systems

AI systems introduce distinctive forms of risk that are subtle, emergent, and often counterintuitive and that arise from the statistical and adaptive nature of learned intelligence. Fragility, distribution shift, hallucination, and the appearance of reasoning without understanding reflect structural limitations rather than engineering defects. These vulnerabilities are intrinsic to systems that rely on historical data, probabilistic inference, and high-dimensional representations of the world.

Managing intelligent systems responsibly therefore requires acknowledging the boundaries of their reliability, anticipating their failure modes, and implementing safeguards that ensure stability and fairness in real-world environments. Continuous monitoring, rigorous validation, transparent governance, and informed human oversight are essential for maintaining trust and preventing the amplification of errors through organizational processes. By understanding how intelligent systems behave, and why they sometimes fail, organizations can deploy AI in ways that enhance decision-making while preserving safety, accountability, and alignment with human values.

CHAPTER 18

Data Risks: Poisoning, Bias, and Integrity Threats

18.1 Introduction: When Data Becomes the Weak Point of Intelligence

Data constitutes the foundational substrate upon which all intelligent systems are built. Because AI models do not perceive or understand the world directly but instead learn from the data supplied to them, the integrity of these systems is inseparable from the integrity of their data. When that data is incomplete, biased, corrupted, or manipulated, the model internalizes those defects. Minor inconsistencies can propagate through a learning system, ultimately producing profound distortions in predictions, automation behavior, and decision outcomes.

Data risks are uniquely insidious because they rarely present themselves as explicit failures. They emerge quietly through mislabeled samples, unnoticed drifts, biased collection practices, subtle corruptions, or maliciously crafted inputs and, subsequently, become embedded within the learned structure of the model.

This chapter examines the multifaceted threats posed by data poisoning, bias, corruption, and provenance failures, and explains how these vulnerabilities compromise the reliability, fairness, and security of AI-driven systems. It also clarifies how weaknesses in data integrity propagate through every stage of the AI lifecycle.

18.2 The Fragility of Data as a Foundation

Unlike program logic, which remains stable until deliberately modified, data is fluid and constantly evolving. Every AI model is a temporal reflection of the data pipeline that supports it. Operational systems introduce noise, sensors deteriorate, human actors make mistakes, and external data feeds change definitions or collection methods. The fragility of AI systems is rooted not only in their models but in the dynamic and imperfect nature of the data that sustains them.

This fragility becomes particularly acute in systems that rely on continuous learning. In such environments, even small perturbations can accumulate over time, creating compounding distortions. When flawed data becomes part of the update process, the model adapts not to reality but to error, resulting in cascading drift. As organizations increase reliance on automation, data must be safeguarded with the same rigor as financial ledgers, operational controls, or critical infrastructure components.

18.3 Data Bias: The Silent Distortion

Bias is one of the most pervasive data risks, emerging long before modeling begins and rooted in structural inequities, historical imbalances, sampling decisions, and organizational practices that unintentionally encode unequal representations of the world. It can appear through underrepresentation or overrepresentation of certain populations, omission of key contextual variables, or the presence of proxy attributes that indirectly encode sensitive characteristics such as race, gender, age, or socioeconomic status. These distortions shape the very foundation on which models are built.

The consequences of bias extend far beyond technical performance. Bias can cause models to serve some groups accurately while failing others, and it can reinforce historical inequities by embedding past patterns into automated decision-making. Hiring algorithms may disadvantage applicants who resemble previously excluded groups, and credit models may perpetuate discriminatory lending patterns. Addressing bias therefore requires more than algorithmic fixes. It demands contextual interpretation, domain expertise, and a deliberate awareness of the social, ethical, and regulatory forces reflected in the data itself.

18.4 Labeling Risks: When Ground Truth Is Not True

Supervised learning relies on labeled data, yet labels often carry uncertainty, inconsistency, or subjectivity. Human annotators may misinterpret ambiguous cases, apply different criteria, or make simple errors. In complex fields such as medicine, law, or performance assessment, labels frequently reflect expert judgment rather than objective truth, introducing variability that models cannot distinguish from factual ground truth.

These labeling risks can significantly distort learning, as models treat labels as authoritative signals. A single misdiagnosis can mislead a clinical model, misclassified security events can compromise detection systems, and transcription errors can alter financial predictions. Crowdsourced workflows add additional vulnerabilities, as annotators may rush for efficiency, misread instructions, or allow personal biases to shape their labeling. Even domain experts may disagree on borderline cases. For these reasons, label quality must be recognized as a core determinant of model reliability, not a peripheral step in data preparation.

18.5 Data Poisoning: When Adversaries Attack the Learning Process

Data poisoning is one of the most severe and stealthy security threats facing AI systems. In such attacks, adversaries intentionally manipulate training data to distort the model's behavior by altering labels, inserting malicious samples, or embedding hidden triggers. Because machine learning systems inherently trust their training data, poisoning can corrupt a model's outputs without altering its code, architecture, or infrastructure, making it especially difficult to detect.

These attacks can take many forms. Fraudsters may inject fabricated transactions to avoid detection, malicious actors may corrupt spam or content-moderation datasets to allow harmful material through, and hackers may tamper with user-generated data to manipulate sentiment analysis or recommendations. Many poisoning attacks embed backdoors, causing the model to behave normally until a specific trigger prompts malicious behavior. In high-stakes applications, such hidden compromises can lead to cascading failures, making it essential to secure training pipelines and rigorously validate data integrity.

18.6 Data Corruption and Pipeline Failures

Not all data risks arise from adversarial intent. Many stem from mundane operational failures such as malfunctioning sensors, broken integrations, schema changes, improperly formatted inputs, or data lost during transmission. These failures can be as damaging as active attacks. A single corrupted field may destabilize a predictive model, a system update may silently break compatibility, and a backlog in logging infrastructure may result in incomplete datasets.

Unlike traditional software systems, where errors typically produce explicit failures, corrupted data can generate plausible but incorrect predictions. This makes data corruption especially dangerous as the system appears functional even as its outputs drift from reliability. Mitigating these risks requires rigorous input validation, schema enforcement, redundancy, and continuous monitoring of data quality across the entire pipeline.

18.7 Adversarial Manipulation of Inputs

Adversarial manipulation refers to targeted attempts to deceive models at inference time by exploiting their sensitivity to subtle perturbations. Attackers craft inputs that appear normal to humans but trigger incorrect classifications or decisions, revealing the fragility of high-dimensional model boundaries. These manipulations can occur across modalities such as images altered at the pixel level to evade vision systems, subtly perturbed text sequences that mislead language models, or manipulated sensor signals that distort real-time analytics.

As AI systems increasingly interact with physical and operational environments, these threats become more consequential, affecting autonomous vehicles, robotics, industrial machinery, and diagnostic tools. Mitigating adversarial risks therefore requires a combination of architectural robustness, adversarial training, input sanitization, confidence calibration, and real-time anomaly detection. The susceptibility of models to small, imperceptible perturbations must be recognized as a structural vulnerability requiring continuous oversight.

18.8 Missing Data and Selection Bias

Missing data presents a significant challenge because it rarely occurs at random. It often reflects systemic inequities, behavioral patterns, accessibility barriers, or operational breakdowns. When certain groups, events, or conditions are consistently underrepresented, models learn incomplete or distorted views of reality. These gaps weaken generalization and can lead to misleading or discriminatory predictions.

Selection bias further compounds the issue when the training dataset does not accurately reflect the population where the model will be deployed. Diagnostic models trained on one demographic may perform poorly on others, and credit systems built on traditional applicants may fail in contexts involving unbanked or underbanked populations. Addressing these risks requires understanding the causes of missing data, ensuring representative sampling, and designing systems that explicitly compensate for structural gaps in the data.

18.9 Data Provenance, Security, and Lineage

As datasets grow in scale and complexity, ensuring trustworthy provenance becomes crucial. Data acquired from external sources may include hidden biases, undocumented transformations, licensing constraints, or embedded errors. Without traceability, organizations cannot verify whether the data aligns with ethical standards, regulatory requirements, or operational needs.

Weak provenance undermines fairness, legality, security, and reproducibility. It impedes audits, inhibits error diagnosis, and compromises the ability to explain or justify model behavior. Maintaining detailed lineage, including tracking the origin, transformation, and usage of data, is essential for accountability and for building models that can be trusted in high-stakes environments.

18.10 Conclusion: Protecting the Integrity of the Intelligence Pipeline

Data risks undermine the foundation of intelligent systems in ways that are subtle, pervasive, and often difficult to detect. Poisoning attacks corrupt learning, bias distorts outcomes, corruption destabilizes predictions, missing data produces blind spots, adversarial inputs exploit model fragility, and weak provenance erodes trust and accountability. These vulnerabilities are not peripheral concerns. They are central determinants of AI reliability, fairness, and safety.

Protecting the integrity of the data pipeline requires robust governance, resilient technical infrastructure, careful human oversight, and continuous monitoring. As organizations depend increasingly on intelligent systems to guide critical decisions, safeguarding data quality and security becomes indispensable to responsible AI practice.

CHAPTER 19

Data Security in AI-Driven Environments

19.1 Introduction: Security in a World of Intelligent Systems

As artificial intelligence becomes deeply embedded within organizational infrastructures, data security shifts from a technical requirement to a foundational determinant of institutional trust and operational resilience. AI systems depend on continuous flows of sensitive, high-value information such as personal, financial, behavioral, operational, and strategic data. These systems expand the number of entities that process data, diversify the environments in which it moves, and increase the frequency with which it is transformed. With this expansion comes a corresponding escalation in risk. The same data that enables intelligence also becomes an attractive target for adversaries seeking to disrupt operations, exfiltrate sensitive information, corrupt models, or exploit system vulnerabilities.

Traditional security frameworks, centered on protecting databases, servers, and perimeter networks, are insufficient in AI contexts. Modern security must protect data throughout its lifecycle including collection, transmission, storage, preprocessing, model training, inference, and archival. It must account for threats that arise not only from external actors but also from the architectural properties of AI systems themselves, such as model inversion, data leakage, adversarial manipulation, and tampering during iterative retraining.

This chapter explores security risks in AI systems, including how threats to data and models reshape traditional cybersecurity and require new protections across the AI lifecycle. It examines how adversarial attacks, data corruption, and model exploitation challenge the integrity and reliability of intelligent systems.

19.2 The Expanding Attack Surface in AI Ecosystems

AI fundamentally expands the organizational attack surface. Data travels across complex, distributed pipelines; resides in heterogeneous storage systems; and interacts with models whose internal representations may embed sensitive patterns. Continuous ingestion from external sources increases exposure to manipulation. Cloud-based training environments, API-driven model services, and real-time streaming architectures introduce additional vectors for adversarial entry.

The AI model itself becomes a security asset and a security liability. It can be probed, reverse-engineered, or misled. It can reveal latent information about training data or be distorted by carefully crafted input sequences. Moreover, attackers increasingly exploit the opacity and complexity of AI systems, using them both as tools for attack and as targets. The scale and dynamism of modern AI ecosystems require security frameworks that are adaptive, multilayered, and deeply integrated into the data and model lifecycle.

19.3 Protecting Data During Collection and Ingestion

The collection and ingestion phase represents the first and often most vulnerable step in the AI lifecycle. Data enters the system from diverse sources, such as user interactions, sensors, enterprise applications, logs, and third-party platforms. It frequently includes sensitive attributes. If these channels are not securely authenticated, encrypted, and validated, attackers may intercept data, manipulate streams, introduce malicious payloads, or compromise authentication tokens.

AI-specific risks further amplify exposure. Because ingestion pipelines often operate autonomously and continuously, adversaries can introduce small but persistent manipulations, poisoning attacks, that shape model behavior while evading detection. Securing ingestion requires robust encryption in transit, source authentication, integrity verification, and real-time anomaly detection capable of identifying patterns indicative of systemic manipulation.

19.4 Securing Data in Storage: From Raw Lakes to Feature Stores

Storage systems in AI environments are heterogeneous and multilevel. Data lakes retain raw, unprocessed information, warehouses curate analytical datasets, lakehouses unify structured and unstructured assets, and feature stores maintain engineered representations used during training and inference. Each layer presents distinct security obligations.

Unstructured data—images, audio, video, logs, text, embeddings—often receives weaker protection despite its sensitivity. Such assets can reveal personal identities, behavioral patterns, confidential communications, or proprietary strategies. Therefore, security must adopt a comprehensive approach, which includes encryption at rest, strict identity- and role-based access controls, granular authorization controls, audit logging, and continuous monitoring.

Protecting stored data is essential not only for privacy but also for ensuring that models do not learn from compromised or unauthorized sources.

19.5 Securing Data in Motion: The Vulnerability of Pipelines

Data pipelines connect ingestion to storage, storage to training, and training to deployment. These flows create transient representations and intermediate artifacts that, if not secured, expose sensitive information. Because pipelines often span cloud environments, distributed compute clusters, and third-party systems, traditional perimeter defenses become inadequate.

Modern pipeline security requires authenticated and encrypted API communication, signed transformations to maintain integrity, strict validation of intermediate states, and continuous monitoring for anomalies in data flow. Ensuring the security of data in motion is essential, as compromised pipelines introduce subtle distortions that propagate throughout the learning ecosystem.

19.6 The Unique Challenge of Protecting AI Models

AI models are themselves high-value assets. A trained model embodies intellectual property, encapsulates sensitive features of training data, and forms the operational core of many decision systems. Model theft can provide adversaries with strategic insights, while model tampering can introduce harmful behaviors that are difficult to detect.

Threats include model inversion (reconstructing sensitive data from outputs), membership inference (determining whether specific records were part of the training set), adversarial attacks, poisoning, and exploitation of model APIs. Protecting models requires controlled access, rate limiting, adversarial testing, integrity verification, encrypted storage, and robust monitoring for suspicious usage patterns. AI security therefore extends beyond data protection to encompass the confidentiality, integrity, and resilience of the models themselves.

19.7 Securing the Model Training Process

Training environments, particularly in cloud and distributed infrastructures, introduce complex security challenges. These processes rely on large-scale datasets, high-performance compute clusters, and extensive parallelization. Misconfiguration, inadequate isolation, or dependency vulnerabilities can expose training pipelines to unauthorized access or manipulation.

Attackers who infiltrate training workflows can subtly alter datasets, embed backdoors, modify model parameters, or inject malicious objectives. Because such alterations often appear benign during initial testing, they may persist undetected into production. Securing training requires strict environment isolation, controlled access to compute resources, integrity checks for datasets and model artifacts, and continuous monitoring throughout the training lifecycle.

19.8 Protecting Data and Models During Deployment

Deployment transforms AI models into operational assets accessible through applications, APIs, and edge devices. This phase introduces new vectors for attack, including unauthorized querying, adversarial probing, inference-time manipulation, and exploitation via unsecured endpoints. Edge deployments pose additional challenges, as physical hardware may fall outside organizational control.

Effective deployment security requires authentication, input validation, encryption, rate limiting, runtime anomaly detection, and mechanisms that prevent the extraction of model parameters or sensitive embedded patterns. Because deployed systems directly influence decisions, failures at this stage can manifest immediately and at scale.

19.9 Ensuring Confidentiality and Preventing Data Leakage

AI systems may inadvertently reveal sensitive information through their outputs. Leakage can occur when models replicate memorized training examples, infer protected attributes, or disclose statistical details that allow reconstruction attacks. Generative models pose particular risks because fluent outputs create the illusion of safe communication even when underlying data is exposed.

Mitigation strategies include privacy-preserving training techniques, output auditing, post-processing constraints, minimization of model overfitting, and formal privacy guarantees. Safeguarding confidentiality requires recognizing that the boundary between data and model is porous. Therefore, models must be governed as potential sources of disclosure.

19.10 The Role of Monitoring and Incident Response in AI Security

Because threats evolve and AI systems change over time, monitoring is essential for detecting anomalies, unauthorized access, model manipulation, or suspicious query patterns. Incident response plans must be tailored to AI-specific scenarios, including compromised training data, corrupted models, poisoning attempts, and emergent behaviors arising from adversarial exploitation.

Effective security requires interdisciplinary coordination among security engineers, data scientists, data engineers, and operational leaders. Rapid detection and mitigation are critical, as delays can lead to cascading failures across AI-dependent processes.

19.11 AI Security as a Cultural and Strategic Imperative

AI security cannot be achieved through technology alone; it must be woven into the fabric of organizational culture. This requires governance structures that prioritize security from the moment data is collected to the point intelligence is deployed. Leadership must set clear expectations, and employees must receive ongoing training that frames security as a shared responsibility rather than a technical burden. When organizations internalize security as a core value, they create an environment where risks are identified early, safeguards are consistently applied, and vulnerabilities are addressed proactively rather than reactively.

As AI systems increasingly influence financial decisions, clinical recommendations, supply chain operations, public services, and risk assessments, security becomes a strategic necessity rather than an operational detail. The integrity of these systems directly affects institutional credibility and public confidence. A single breach, poisoning event, or adversarial manipulation can undermine trust and produce cascading consequences across interconnected domains. Treating AI security as a cultural and strategic imperative ensures that organizations not only protect their systems but also preserve the trust of the people and institutions that depend on them.

19.12 Conclusion: Security as the Backbone of Trustworthy AI

Data security enables the stability, fairness, and reliability of AI-driven systems. Securing data, models, pipelines, and outputs throughout the lifecycle ensures that intelligence supports organizational goals without exposing sensitive information or creating new vulnerabilities. In environments where AI influences critical decisions, security becomes a foundation for trust, accountability, and long-term resilience. Therefore, protecting the integrity of the intelligence pipeline is not merely a defensive obligation but a strategic imperative for responsible AI deployment.

CHAPTER 20

Organizational Risks: Misuse, Automation Bias, and Overreliance

20.1 Introduction: When the Human-Machine Relationship Becomes the Risk

Artificial intelligence transforms organizations not only through its computational capabilities but through the ways humans perceive, interpret, and depend upon it. While technical vulnerabilities such as data corruption, adversarial interference, drift, and model fragility are vital concerns, the most persistent risks arise from organizational behavior, institutional norms, and cognitive tendencies. Intelligent systems are embedded within human processes, and their effectiveness depends on how individuals and teams design, deploy, and respond to them. When this relationship becomes unbalanced, AI shifts from a strategic asset to a catalyst of systemic vulnerability.

Organizational risks emerge when AI is misapplied, misunderstood, or relied upon without sufficient oversight. These risks stem not from algorithmic architecture but from the structural, cultural, and psychological dynamics of the enterprise. Misuse can distort decisions, automation bias can magnify errors, and overreliance can erode human judgment. As intelligent systems become integral to operations, understanding the human-machine relationship becomes essential for sustaining responsible and resilient organizational performance.

This chapter examines the human and organizational risks that emerge when AI is misunderstood, misused, or overly trusted. It explores how automation bias, skill erosion, and blind reliance can quietly undermine decision quality and institutional resilience. It also highlights how governance gaps and cultural complacency amplify these risks over time.

20.2 Misuse: When AI Is Applied to Problems It Was Never Designed to Solve

Misuse occurs when AI is deployed in contexts that exceed its intended scope or violate its underlying assumptions. Organizations often overextend models due to enthusiasm for automation, institutional pressure to adopt AI broadly, or misunderstandings about system capabilities. A tool designed for customer service may be repurposed for hiring decisions, or a maintenance model may be applied to financial forecasting, simply because both appear "predictive." Such repurposing disregards the contextual and ethical constraints that shaped the original design.

The danger lies in deploying models where data is insufficient, patterns are unstable, or domain knowledge is limited or lacking. Misuse produces errors that are difficult to diagnose, introduces inequities, and creates false assurances of analytical rigor. Preventing misuse requires governance structures that define model intent clearly, maintain documentation, and facilitate dialogue between technical teams and domain experts. When oversight is absent, AI can amplify poor decisions rather than improve them.

20.3 Misinterpretation: When Teams Overread AI Outputs

AI systems generate predictions and probabilistic patterns, not definitive truths. Yet within organizations, these outputs are often interpreted as authoritative judgments. Probabilities are mistaken for certainties, correlations are interpreted as causal relationships, and confidence scores are assumed to represent reliability rather than model-internal assessments. Misinterpretation arises because AI outputs often appear polished or linguistically fluent, creating an illusion of clarity and certainty.

This risk intensifies when users lack statistical literacy or when systems present outputs without adequate explanation of uncertainty. A risk score may be construed as a diagnosis, a forecast as a guarantee, and a recommendation as an operational directive. Such misinterpretations narrow the space for human judgment and introduce systematic error into decision-making processes. Mitigation requires clear communication of uncertainty, appropriate calibration of outputs, and organizational investment in AI literacy so that predictions inform decisions rather than dictate them.

20.4 Automation Bias: The Human Tendency to Rely Too Heavily on Intelligent Systems

Automation bias arises when individuals defer to AI-generated outputs even when contradicting evidence or expertise is available. Modern systems produce fluent language, sophisticated visualizations, and rapid analysis, creating a perception of authority that can overshadow human intuition. When confronted with conflicting interpretations, individuals may privilege the machine's answer merely because it is computational.

This bias can erode safety in high-stakes contexts. Clinicians may over-rely on diagnostic scores, financial analysts may defer to algorithmic forecasts, and compliance officers may accept model-generated risk assessments without critical evaluation. Over time, human oversight diminishes as the system becomes normalized within workflows. Preventing automation bias requires fostering a culture of critical engagement, embedding explicit uncertainty indicators, and designing interfaces that encourage users to maintain evaluative agency.

20.5 Overreliance: When AI Becomes a Crutch Instead of a Tool

Overreliance occurs when organizations treat AI as a replacement for expertise rather than as a complement to it. As systems grow more sophisticated and deeply integrated, individuals may offload essential reasoning, situational awareness, and contextual interpretation to automated processes. This dynamic leads to erosion of skills, diminished human vigilance, and increased vulnerability to unexpected failures.

Operational risks intensify when organizations depend excessively on single models or pipelines. A disruption in a demand forecasting model may halt supply chains, and an error in a credit assessment model may affect entire portfolios. Overreliance centralizes risk and creates brittle decision environments. Maintaining resilience requires ensuring redundancy, establishing alternative workflows, and preserving human discretion in critical processes. AI must serve as an augmentation of human capability, not a substitute.

20.6 Decision Paralysis: When AI Creates Too Many Options or Excessive Complexity

Although AI is often deployed to simplify decision-making, it can also overwhelm users by generating excessive predictions, scenarios, or recommendations. When systems present multiple outputs, each accompanied by confidence intervals or narrative explanations, individuals may struggle to determine which path to follow. Instead of resolving uncertainty, AI introduces complexity that inhibits timely action.

Decision paralysis emerges when organizations assume that every insight must be considered or that every model output carries equal weight. The result is indecision, fragmentation of responsibility, or reliance on arbitrary heuristics. Addressing this risk requires deliberate interface design, prioritization mechanisms, and decision frameworks that reduce cognitive load. AI must guide decisions by structuring information, not by inundating users with unfiltered possibilities.

20.7 Organizational Blind Spots: When AI Reinforces Existing Structures

AI systems often encode the incentives, worldviews, and historical patterns embedded within the organizations that deploy them. When developed without diverse input or broad stakeholder engagement, AI may reinforce existing inequalities, institutional assumptions, or internal power structures. This produces organizational blind spots, areas where risks remain unnoticed because the system reflects only a narrow perspective.

Blind spots can undermine strategic decisions. Models may privilege quantitative metrics at the expense of qualitative insights, optimize for efficiency without considering social or ethical implications, or lock the organization into outdated operational frameworks. These vulnerabilities remain hidden until they manifest as public failures or internal conflict. Addressing them requires inclusive design practices, cross-functional review processes, and governance mechanisms that ensure AI aligns with the organization's broader mission and values.

20.8 The Organizational Cost of "AI Hype" and Unrealistic Expectations

AI-related enthusiasm often leads organizations to adopt inflated expectations about the technology's transformative capacity. Leaders may anticipate rapid efficiency gains, employees may fear displacement, and project teams may rush to deploy systems without adequate evaluation. When expectations outpace capabilities, organizations experience misallocated resources, stalled initiatives, and declining trust.

Hype creates strategic distortions as teams pursue ill-defined use cases, treat AI as a solution in search of a problem, or neglect foundational data and process improvements. Beyond financial waste, unrealistic expectations damage organizational cohesion, foster skepticism, and undermine long-term adoption. Sustainable AI integration requires tempering ambition with critical realism and grounding initiatives in demonstrable value rather than aspirational narratives.

20.9 Organizational Drift: When AI Slowly Alters Behaviors, Norms, and Values

AI systems influence how decisions are made, how performance is evaluated, and how individuals interact with one another. Over time, these influences can reshape organizational culture. Metrics-driven models may incentivize employees to optimize for measurable indicators rather than substantive goals. Automated workflows may standardize decisions that require human nuance, reducing discretion, empathy, or professional judgment.

Such drift occurs gradually, often without deliberate oversight. As AI-mediated practices become normalized, they may misalign with the organization's values or mission. Understanding these cultural consequences requires continuous reflection, structured feedback mechanisms, and governance processes that evaluate how AI shapes behavior. Responsible leadership ensures that cultural evolution enhances, rather than erodes, institutional integrity and human-centered values.

20.10 Conclusion: Building a Human-Centered Approach to AI

Organizational risks arise not from the intrinsic properties of intelligent systems but from the ways people and institutions engage with them. Misuse, misinterpretation, automation bias, overreliance, blind spots, and cultural drift emerge when AI is introduced without sufficient governance, contextual understanding, or critical oversight. A resilient organization recognizes that AI must augment human judgment, not override it.

A human-centered approach to AI emphasizes clarity of purpose, thoughtful design, cross-functional collaboration, and sustained attention to how systems influence behavior over time. By aligning AI with organizational values and empowering individuals to engage with technology responsibly, organizations can harness the benefits of intelligent systems while mitigating the socio-technical risks that accompany them.

CHAPTER 21

Compliance, Regulation, and Accountability in AI Systems

21.1 Introduction: The Era of Regulated Intelligence

As artificial intelligence becomes a foundational element of organizational operations, its capabilities increasingly intersect with legal, ethical, and societal mandates. No longer viewed as an experimental or exploratory technology, AI now influences medical diagnoses, financial decisions, hiring practices, public services, and national security. With this expanded influence comes the expectation that AI systems must be accountable, fair, transparent, and safe. Governments, regulatory bodies, and international standards organizations are responding by defining rules that guide the development, deployment, and oversight of intelligent systems. Therefore, compliance is not an optional enhancement; it is a strategic and legal necessity.

This chapter explores the regulatory and compliance frameworks governing AI systems, including the EU AI Act, NIST guidance, ISO standards, and sector-specific rules. It examines how accountability and compliance shape the design, deployment, and oversight of intelligent systems. It also clarifies how regulatory alignment directly influences organizational risk, trust, and long-term AI adoption.

21.2 Why Regulation Matters: Aligning AI With Societal Expectations

AI systems operate within environments shaped by deeply held norms related to fairness, privacy, autonomy, equality, and due process. When unregulated, these systems can amplify historical injustices, erode privacy, or generate harmful decisions with limited avenues for recourse. Regulation exists to ensure that the power of AI aligns with societal expectations rather than displacing them.

Regulation also provides consistency and predictability in global markets. Organizations deploying AI across borders face varying legal obligations, and harmonized frameworks reduce uncertainty while supporting scalable compliance. Most importantly, regulation reinforces accountability, as it requires organizations to document their decisions, evaluate risks, and justify the design and behavior of systems that affect individuals and communities.

21.3 The EU AI Act: A Risk-Based Regulatory Framework

The European Union's AI Act represents the most ambitious and comprehensive attempt to establish legally binding standards for AI. Its risk-based approach recognizes that not all AI systems pose the same threats. Systems classified as unacceptable risk, such as manipulative or discriminatory surveillance, are prohibited outright. High-risk systems, including those used in healthcare, employment, education, infrastructure, and public safety, must meet stringent requirements concerning transparency, data governance, robustness, documentation, and human oversight. Lower-risk systems may require disclosure or labeling, particularly when interacting directly with consumers.

The EU AI Act emphasizes process governance. It requires evidence of responsible practices across the entire lifecycle, including how data was sourced, how models were trained, how risks were mitigated, and how oversight is operationalized. This regulatory model elevates AI governance from a discretionary organizational activity to a mandatory discipline supported by rigorous documentation and auditability.

21.4 NIST's AI Risk Management Framework: Managing AI as a Socio-Technical System

In the United States, NIST has developed the AI Risk Management Framework (AI RMF), which conceptualizes AI not as a purely technical construct but as a socio-technical system. This perspective recognizes that intelligent systems interact with people, institutions, and social structures, and that the associated risks cannot be mitigated through engineering alone.

The NIST framework, though voluntary, guides organizations in assessing risks, documenting model behavior, enhancing transparency, stress-testing systems, and incorporating human factors into design and oversight. It emphasizes reliability, fairness, accountability, and privacy as cornerstones of responsible deployment. As regulators, auditors, and industry groups increasingly reference the NIST framework, it is becoming a de facto standard for AI governance in the United States.

21.5 ISO Standards: Formalizing the Discipline of AI Governance

International standards bodies, particularly the International Organization for Standardization (ISO), have begun developing comprehensive frameworks that formalize the practice of AI governance. Among the most significant are ISO/IEC 42001, which establishes requirements for AI Management Systems, ISO/IEC 23894, which provides guidelines for AI risk management, and ISO/IEC 22989, which defines foundational concepts and terminology for artificial intelligence. Together, these standards offer structured, certifiable pathways for organizations seeking to institutionalize responsible AI practices.

These frameworks transform AI oversight from an ad hoc organizational effort into a systematic discipline characterized by rigorous documentation, consistent validation, defined quality measures, and transparent governance structures. They guide organizations in embedding ethical principles, regulatory requirements, and operational safeguards into the entire lifecycle of an AI system. By establishing common expectations and harmonized processes, ISO standards promote global interoperability, enabling organizations to deploy intelligent systems safely and reliably across diverse jurisdictions, partners, and industries.

21.6 Sector-Specific Regulations: When AI Determines High-Stakes Outcomes

AI often operates within sectors already governed by extensive regulatory frameworks. Financial institutions must comply with anti-discrimination laws, model risk management rules, and auditability requirements. Healthcare organizations must protect patient data under HIPAA, adhere to clinical safety standards, and validate the performance of diagnostic algorithms. Employers deploying hiring or performance evaluation tools must meet equal employment opportunity regulations. Public-sector agencies must justify automated decisions that affect rights, access, or benefits.

Sector-specific regulations recognize that intelligent systems influence core dimensions of human well-being. They require documentation of how decisions are made, safeguards against discrimination, and channels for individuals to challenge or appeal automated outcomes. Therefore, compliance becomes a multidisciplinary responsibility, involving not only data scientists but also legal experts, ethicists, and domain professionals.

21.7 Accountability: Ensuring Someone Owns the Decisions AI Makes

Accountability is the organizing principle that underpins all AI regulation. When a system influences a decision, someone must be responsible for its consequences. Accountability requires organizations to designate clear ownership for model development, deployment, monitoring, and risk management. It demands documentation of assumptions, design choices, data sources, and operational behavior.

Accountability also extends to external transparency. Individuals affected by automated decisions must be informed about the system's role, the basis for decisions, and available avenues for appeal or review. Accountability transforms AI from a black-box technology into a governed system with clearly delineated responsibilities and expectations.

21.8 Documentation and Traceability: The Audit Trail of Intelligence

Documentation is a foundational requirement of trustworthy AI. Regulators expect organizations to maintain detailed records of model lineage, data provenance, preprocessing steps, feature engineering, training procedures, risk assessments, testing results, and oversight mechanisms. Traceability ensures that organizations can reconstruct the reasoning behind model decisions, diagnose failures, and demonstrate compliance to auditors or regulators.

Beyond compliance, documentation supports operational resilience. When models drift, require retraining, or must be redesigned, an accurate audit trail allows teams to understand earlier choices and replicate essential steps. In high-stakes domains, documentation is not an administrative burden; it is an essential safeguard.

21.9 Human Oversight: The Critical Safeguard Against Automated Harm

Regulators increasingly mandate that high-impact AI systems include meaningful human oversight. This requirement acknowledges that even robust models cannot replace contextual judgment, ethical reasoning, or empathy. Human oversight ensures that decisions with significant consequences, such as medical diagnoses, credit denials, hiring outcomes, or risk assessments, are reviewed by individuals who can interpret ambiguity and intervene when needed.

Effective oversight requires more than nominal human participation. Humans must be empowered with authority, training, and visibility. Interfaces must communicate uncertainty and rationale. Escalation pathways must exist for ambiguous or high-risk cases. Without these conditions, oversight becomes symbolic rather than substantive.

21.10 Conclusion: Governance as the Foundation of Responsible AI

Compliance and regulation define the expectations for responsible AI by establishing standards for fairness, transparency, safety, and accountability. These frameworks do not impede innovation as they ensure that innovation advances in ways that protect rights, reinforce trust, and align with democratic and ethical values. Effective AI governance requires interdisciplinary collaboration, vigilant monitoring, rigorous documentation, and a commitment to ensuring that intelligent systems serve societal and organizational goals responsibly.

CHAPTER 22

Compute Infrastructure: GPUs, Cloud, Edge, and Distributed AI

22.1 Introduction: The Computational Backbone of Modern Intelligence

The rise of artificial intelligence has fundamentally reshaped how organizations conceive of compute infrastructure. Earlier eras of analytics required limited computational power, as they centered on statistical modeling, business reporting, and modest-scale data processing. Modern AI, particularly deep learning and generative architectures, demands significantly more. Training advanced neural networks requires massive parallelism, high-bandwidth data movement, and specialized accelerators capable of performing billions of operations per second. Deploying these systems reliably at scale further requires infrastructure that is fast, resilient, flexible, and tightly integrated into organizational workflows.

Compute infrastructure is not a peripheral technical concern, as it is the foundational enabler of intelligent systems. Its architecture, spanning GPUs, cloud platforms, edge devices, and distributed computing, determines what systems can be built, how quickly teams can iterate, and how effectively organizations can operationalize AI.

This chapter explores the computational foundations of modern AI, including how compute infrastructure, cloud platforms, and specialized hardware enable scalable, high-performance, and reliable intelligence. It also explains why compute capacity has become a strategic constraint on AI innovation and deployment.

22.2 Why AI Requires Specialized Computing

AI workloads differ fundamentally from traditional computational tasks. Deep learning models rely heavily on matrix multiplications and tensor operations that benefit from highly parallel execution. While CPUs excel at general-purpose, sequential computation, they cannot match the throughput needed for large-scale training. GPUs, originally designed for rendering graphics, emerged as the ideal platform because their architecture, thousands of small cores optimized for simultaneous processing, aligns precisely with AI's numerical demands.

As models grow more complex and training datasets expand, compute requirements scale sharply. Training a modern language model or multimodal architecture may require weeks of continuous computation across clusters of GPUs. Even inference, producing predictions or generating language, can be resource-intensive, especially for real-time applications or high-volume workloads. Specialized accelerators such as TPUs and custom AI chips further extend these capabilities, enabling efficiency gains but also introducing new technological dependencies. The evolution of computing infrastructure thus shapes not only AI's frontier but also the economic and operational landscape of AI development.

22.3 GPUs: The Workhorses of Modern AI

GPUs have become the primary platform for modern machine learning because of their exceptional capacity to perform parallel numerical operations. Neural network training depends on repeatedly multiplying large matrices, a task GPUs can execute concurrently at massive scale, reducing training time dramatically compared with CPU-based approaches. Their architectures also support advanced optimization methods, such as mixed-precision training, that further accelerate computation while preserving accuracy. For enterprises, clusters of GPUs enable rapid experimentation and faster iteration cycles, expanding the range of problems that can be tackled efficiently.

The adoption of GPUs, however, introduces significant operational considerations. They require specialized programming frameworks, careful job scheduling, extensive cooling, and substantial energy resources. Organizations must weigh these demands against performance gains, ensuring infrastructure is both cost-effective and sustainable. Despite these challenges, GPUs remain the cornerstone of AI compute strategy due to their flexibility, strong ecosystem support, and consistently high performance across a wide range of machine learning workloads.

22.4 Cloud Computing: Scaling AI Without Physical Constraints

Cloud computing has transformed AI development by enabling organizations to provision GPU clusters, distributed training environments, and high-performance storage on demand. This elasticity allows teams to scale resources during intensive training cycles and release them afterward, improving cost efficiency and accelerating experimentation. Cloud platforms also support global collaboration, reduce the burden of managing physical infrastructure, and provide rapid access to cutting-edge hardware that would be costly or impractical to maintain in-house. Hybrid cloud models extend this flexibility by allowing sensitive data to remain on-premises while leveraging cloud capacity for large-scale computation, balancing regulatory requirements with scalability.

Despite these advantages, cloud adoption brings its own set of challenges. Usage-based pricing can lead to unpredictable costs, particularly during large training runs, latency can limit the viability of real-time applications, and dependence on external providers introduces strategic and operational risks. Building an effective cloud strategy therefore requires thoughtful governance, careful workload planning, and ongoing evaluation of performance-to-cost trade-offs to ensure that cloud resources are used efficiently and sustainably.

22.5 Edge Computing: Bringing Intelligence Closer to the Source

In many domains, intelligence must operate near the point of data generation rather than in remote data centers. Edge computing enables real-time inference in environments with strict latency requirements, limited connectivity, or privacy constraints. Autonomous vehicles, industrial control systems, healthcare devices, and IoT platforms depend on edge intelligence to make rapid, context-aware decisions.

Edge devices face practical limitations, such as restricted compute power, energy constraints, and limited storage. As a result, models must be optimized through compression, precision scaling, pruning, or architectural redesign. Despite these limitations, edge computing enhances resilience by enabling continued operation during network disruptions and reduces privacy exposure by processing sensitive data locally. The interplay between cloud and edge systems creates hybrid intelligence, where complex training occurs centrally and streamlined inference occurs locally.

22.6 Distributed Computing: Coordinating Intelligence Across Many Machines

As datasets and models expand, single-machine computation becomes insufficient. Distributed computing addresses this need by partitioning workloads across multiple nodes that collaborate during training or inference. Distributed training shortens development cycles for large models and enables architectures that would otherwise exceed the memory or compute limits of individual machines.

Distributed computing also supports large-scale data processing, essential for domains with continuous or high-volume data generation such as telecommunications, logistics, e-commerce, and scientific research. These systems introduce architectural challenges including synchronization, communication overhead, and fault tolerance. However, when properly orchestrated, they enable AI systems to operate at industrial scale.

22.7 Energy, Sustainability, and the Cost of Intelligence

AI's computational intensity has raised significant questions about energy consumption and environmental impact. Training sophisticated models requires substantial electricity and generates heat that necessitates advanced cooling systems. Cloud data centers place pressure on regional energy infrastructure, and large-scale deployments contribute to growing carbon footprints.

Responsible AI development requires evaluating not only accuracy and performance but also environmental cost. Efficiency techniques, such as hardware acceleration, model pruning, precision scaling, and optimized training regimes, help reduce compute demands. Cloud providers increasingly offer renewable-energy options or carbon-neutral data centers. Organizations must carefully weigh the benefits of highly complex models against their environmental impact, integrating sustainability into long-term AI strategies.

22.8 Reliability, Redundancy, and Fault Tolerance in AI Infrastructure

AI systems often support mission-critical operations, making infrastructure reliability essential. Hardware failures, network outages, and software malfunctions can destabilize AI pipelines or interrupt decision-making processes. Redundancy, through backup clusters, alternative pathways, and failover architectures, ensures continuity even when components fail. Fault-tolerant systems detect anomalies, reroute workloads, and maintain degraded but functional performance until full capacity is restored.

These architectural principles transform compute infrastructure into a resilient platform capable of supporting uninterrupted AI operations, even in the face of unexpected disruptions. Organizations can maintain service continuity and prevent localized failures from cascading across the system by incorporating layered safeguards, such as automated failover, real-time anomaly detection, and dynamic workload redistribution. Reliable, redundant, and fault-tolerant designs ultimately ensure that AI systems remain dependable at scale, preserving operational stability when it matters most.

22.9 Choosing the Right Compute Strategy for the Enterprise

There is no universal compute architecture suitable for all organizations. Optimal strategy depends on data volume, latency requirements, regulatory constraints, security considerations, cost structures, and strategic objectives. Many enterprises adopt hybrid approaches that combine cloud scalability, on-premises control, and edge responsiveness.

Compute planning must be cross-functional, involving engineering, data science, security, operations, and finance. Decisions about infrastructure shape innovation velocity, operational resilience, and the long-term viability of AI initiatives. Compute is not merely an IT detail but a strategic asset that underpins organizational intelligence.

22.10 Conclusion: Infrastructure as the Engine of Intelligent Transformation

Compute infrastructure enables the capabilities that define modern AI. GPUs accelerate model development, cloud platforms deliver elasticity, distributed systems support scale, and edge devices enable real-time responsiveness. Together, these elements form the technological backbone of intelligent organizations. The success of AI depends not only on algorithms and data but also on the robustness and strategic design of the systems that support them.

CHAPTER 23

Platforms and Tools: Building Intelligent Analytics Environments

23.1 Introduction: The Emergence of the Intelligent Analytics Stack

As artificial intelligence reshapes the foundation of modern enterprises, organizations must construct environments capable of supporting the full lifecycle of intelligent systems. These environments, which encompass data platforms, machine learning pipelines, orchestration systems, and monitoring frameworks, form the operational layer through which AI generates business value. Without a coherent platform strategy, AI initiatives become fragmented, inconsistent, and difficult to scale. Models developed by one team may be incompatible with another's infrastructure, insights may remain siloed, and pipelines may fail when transferred across environments.

Platforms and tools therefore play a central role in transforming AI from isolated prototypes into enterprise-grade capabilities. They standardize the processes by which data is gathered, transformed, modeled, deployed, and monitored. They ensure reliability, reproducibility, and alignment across teams. Most importantly, they enable organizations to scale intelligently, not by increasing technical labor indefinitely, but by building environments that automate complexity, promote governance, and facilitate interdisciplinary collaboration.

This chapter explores the platform and tool ecosystems that support contemporary AI systems, including how they enable development, deployment, orchestration, and operational scalability across the AI lifecycle. It also highlights how platform choices influence flexibility, governance, and long-term system interoperability.

23.2 Data Platforms: The Foundation for Intelligent Workflows

Every intelligent system begins with data. AI platforms rely on the ability to ingest, store, organize, and process information at scale. Modern data platforms integrate these functions, including storage, ingestion, transformation, governance, lineage, and access, into unified environments that support both operational and analytical workflows. Unlike earlier architectures that separated transactional workloads from analytical tasks, contemporary platforms increasingly provide shared, near-real-time access to data, enabling seamless use across machine learning and decision-support contexts.

Data platforms must handle diverse modalities, such as relational tables, log files, documents, images, audio, sensor signals, and unstructured natural language. They also must support streaming pipelines, ensuring that real-time models receive timely and reliable inputs. Organizations increasingly adopt lakehouse architectures that combine the flexibility of data lakes with the consistency of data warehouses. These platforms serve as the backbone of intelligent workflows, enabling teams to build features, train models, and operationalize insights without reconstructing foundational infrastructure.

23.3 Machine Learning Platforms: The Operating System for AI Development

With a data foundation in place, organizations must provide structured environments for developing, training, and validating models. Machine learning platforms function as the "operating system" for AI development, consolidating tools for exploration, preprocessing, experiment tracking, version management, and deployment.

These platforms bring coherence to what would otherwise be a fragmented ecosystem of notebooks, scripts, and ad hoc workflows. They capture metadata, including hyperparameters, dataset versions, metrics, and model artifacts, ensuring experiment reproducibility and comparability. They enable collaboration across teams, support automated model management, and streamline the transition from experimentation to production. Whether organizations adopt fully managed ecosystems or build custom stacks, the core objective remains the same: establishing a consistent, scalable environment for transforming data into operational intelligence.

23.4 Orchestration Systems: Coordinating Complex AI Pipelines

AI systems depend on pipelines that integrate data ingestion, validation, feature generation, model training or retraining, deployment, and monitoring. These components must operate reliably and in a coordinated manner across heterogeneous tools, systems, and teams. Orchestration platforms provide the structure through which these workflows are scheduled, automated, and governed.

Effective orchestration manages dependencies, executes tasks in the correct sequence, handles failures gracefully, and records lineage across the entire lifecycle. In large enterprises, orchestration becomes indispensable because the AI pipeline spans multiple environments, including data lakes, feature stores, compute clusters, deployment frameworks, and observability tools. Orchestration systems transform these disparate components into coherent, auditable, and scalable operational processes.

23.5 Feature Stores: Institutionalizing the Most Valuable Assets in AI

Features, the structured representations of data used by models, are among the most valuable assets in machine learning. Yet in many organizations they are created informally, embedded in notebooks, and duplicated across teams. Feature stores address this by providing centralized environments where features are defined, validated, documented, governed, and served consistently for both training and inference.

By standardizing feature engineering practices, feature stores reduce redundancy, eliminate mismatches between training and production data, and accelerate development cycles. They allow teams to reuse and extend prior work, transforming organizational knowledge into a shared and continuously improving asset. Feature stores institutionalize feature engineering as a disciplined component of the AI lifecycle rather than an ad hoc, project-specific activity.

23.6 Model Deployment Platforms: Bringing Intelligence into Production

Models deliver value only when they are deployed into applications, workflows, or decision systems. Deployment platforms provide the infrastructure through which predictions are served, via APIs, batch scoring pipelines, streaming engines, or embedded edge environments. They manage versioning, rollback, scaling, health checks, and resource allocation, ensuring reliability under diverse operational demands.

These platforms bridge the gap between experimentation and real-world usage. They support controlled experimentation (e.g., A/B tests, shadow deployments), enforce governance and auditability, and integrate safeguards for sensitive decision contexts. In high-stakes environments, deployment platforms enable human oversight, explanation mechanisms, and risk-managed rollout. By operationalizing models consistently and safely, they transform them from research artifacts into durable components of enterprise systems.

23.7 Monitoring and Observability: Ensuring Reliability After Deployment

Once a model is deployed, its behavior is shaped by evolving data, shifting environments, and changing user interactions. Monitoring and observability platforms provide visibility into this behavior by tracking performance metrics, latency, error rates, data drift, segment-level fairness indicators, and operational anomalies. Observability extends beyond monitoring by exposing internal signals, enabling root-cause analysis and contextual understanding of model behavior.

These systems alert teams when models degrade, when data distributions shift, or when outcomes deviate from expected patterns. They support drift detection, adaptive retraining, and proactive risk mitigation. In environments where AI influences financial, medical, operational, or policy decisions, monitoring is not optional. It is an essential safeguard against silent model failure.

23.8 Collaboration Platforms: Bridging Technical and Business Teams

AI is inherently interdisciplinary. Data scientists, engineers, domain specialists, analysts, and decision-makers all contribute to the lifecycle of intelligent systems. Collaboration platforms, including dashboards, documentation hubs, conversational analytics, and shared repositories, facilitate communication across these groups. They render complex insights accessible, ensuring that stakeholders understand model intent, limitations, and implications.

Such platforms foster alignment, reduce miscommunication, and strengthen AI literacy across the organization. They enable business teams to interact with models meaningfully and ensure that technical teams remain grounded in organizational priorities and constraints. Collaboration tools thus function as the connective tissue between technical innovation and strategic application.

23.9 Integration Frameworks: Connecting AI to Business Processes

AI creates value only when integrated into the workflows where decisions are made. Integration frameworks connect models to operational systems, such as enterprise resource planning (ERP), customer relationship management (CRM), supply chain platforms, scheduling tools, and customer-facing applications. These frameworks standardize communication between AI outputs and business systems, enabling seamless execution of decisions.

Integration requires deep understanding of business processes, decision points, and the role AI should play. It often poses greater challenges than model development. Integration frameworks reduce this complexity by formalizing interfaces, governing data exchanges, and embedding intelligence into organizational processes. They enable AI to function not as a standalone capability but as a structural component of the enterprise.

23.10 Conclusion: Platforms as the Scaffold for Scalable Intelligence

Platforms and tools provide the structural foundation that enables intelligent systems to operate reliably, consistently, and at scale. They unify disparate components, including data, models, pipelines, monitoring, and workflows, into coherent ecosystems that support continuous learning and improvement. Without these platforms, AI remains experimental and fragile; with them, organizations build durable environments capable of supporting long-term innovation.

CHAPTER 24

Enterprise Integration: Connecting AI to Business Processes

24.1 Introduction: When Intelligence Becomes Operational

Artificial intelligence delivers its true value only when it becomes deeply embedded within the operational fabric of an organization. Models, platforms, and algorithms, regardless of their sophistication, remain abstract assets until they are connected to the business processes where decisions are made, actions are taken, and performance outcomes are realized. Therefore, enterprise integration is the discipline of transforming AI from an analytical capability into an operational one, weaving intelligence into workflows, systems, applications, and human activity. It marks the point at which strategic vision becomes operational reality.

Integration, however, is among the most challenging dimensions of AI adoption. Unlike experimentation conducted in controlled environments, operational AI must operate within complex ecosystems shaped by legacy technologies, regulatory constraints, organizational culture, and human expectations. The task is not merely to "plug in" a predictive model. It is to redesign processes so that intelligence flows reliably, transparently, and coherently across the enterprise.

This chapter explores how AI is integrated into enterprise systems, including the architectural, technical, and organizational dimensions that translate intelligence into operational transformation. It also explains how integration determines whether AI remains experimental or becomes embedded in everyday business operations.

24.2 The Role of Workflows in AI-Driven Organizations

Organizations run on workflows, which are structured sequences of steps that govern how tasks are completed and how responsibilities are allocated. These workflows determine how information moves, how decisions are produced, and how resources are coordinated. AI modifies these workflows by automating certain steps, augmenting human judgment, or reshaping entire processes around predictive capabilities. But effective integration requires more than inserting a model into an existing sequence of tasks. It requires understanding how AI changes the workflow itself.

AI-enabled workflows often blur the boundaries between operational and analytical processes. A forecasting model may automatically adjust inventory allocations. A risk-scoring system may trigger human review. A recommendation engine may dynamically adapt user-facing interfaces. For AI to function effectively in these contexts, workflows must be designed so that insights arrive at the right moment, in the appropriate format, and with sufficient context. Integration thus becomes a design discipline, balancing automation with oversight, efficiency with transparency, and speed with reliability.

24.3 APIs and Microservices: The Connective Tissue of Intelligent Systems

At the technical layer, APIs and microservices are the mechanisms by which models interact with applications. They function as the connective tissue that allows intelligence to circulate across the enterprise. A model may be exposed as a service for real-time scoring, a recommendation engine may provide endpoint responses for customer interfaces, and a generative model may power conversational agents through streaming APIs.

Microservice architectures offer flexibility, modularity, and resilience. They enable teams to deploy models independently, update components without interrupting upstream or downstream systems, and scale services based on demand. They allow organizations to combine traditional applications with AI-driven components, creating hybrid systems in which intelligence enhances existing processes without requiring complete system replacement. Microservices thereby support rapid iteration, safe experimentation, and broad integration across the enterprise.

24.4 Event-Driven Systems: Triggering Intelligence at the Moment It Matters

Many business processes depend on reacting to events, such as transactions, sensor readings, user behaviors, or external signals. Event-driven architectures allow AI systems to respond to these moments dynamically and in real-time. When a customer engages with a product page, a propensity model may estimate purchase likelihood. When machinery exhibits anomalous vibrations, an AI system may initiate a maintenance request. When a suspicious pattern emerges, a fraud model may intervene before harm occurs.

Event-driven integration closes the gap between prediction and action, enabling organizations to operate proactively rather than reactively. It supports continuous feedback loops that help models adapt to evolving conditions. Real-time integration requires low latency, high performance, and robust monitoring. However, when executed effectively, it transforms operational processes into intelligent, adaptive systems.

24.5 Embedding AI into Enterprise Applications

For most employees, AI is encountered not through development environments or dashboards but through the applications they use every day, such as CRM platforms, ERP systems, workflow engines, customer service tools, and productivity applications. Embedding AI into these interfaces ensures that insights are delivered directly at the point of decision-making, reducing friction and enabling nontechnical users to benefit from intelligent systems.

Embedding may involve integrating predictive scores into dashboards, incorporating automated suggestions into workflows, or enabling intelligent assistants that guide users through tasks. However, embedding also requires thoughtful human-centered design. Outputs must be interpretable, contextualized, and aligned with users' goals. Effective interface design ensures that AI enhances judgment rather than obscuring responsibility, making enterprise applications more intuitive, efficient, and adaptive.

24.6 Integrating AI with Legacy Systems

Many organizations rely on legacy systems that were not designed with AI or modern data architectures in mind. Integrating AI with these systems requires creative architectural strategies, such as constructing middleware layers that translate between formats, wrapping legacy systems in modern interfaces, or gradually modernizing infrastructure while maintaining operational continuity.

Legacy integration is as much an organizational challenge as a technical one. It requires balancing the need for innovation with the imperative of reliability. Effective approaches begin with understanding which components can be augmented, which require reengineering, and which should be phased out over time. With careful planning, organizations can introduce intelligence into legacy environments without compromising stability or business continuity.

24.7 Orchestration Across the Enterprise: Ensuring Consistency and Governance

Enterprise integration requires consistent management of workflows across departments, datasets, and platforms. AI systems depend on standardized procedures for data ingestion, feature generation, model deployment, monitoring, and auditing. Orchestration platforms coordinate these processes across the organization, ensuring that models operate consistently and that updates propagate safely.

Orchestration also reinforces governance. As AI is deployed across finance, marketing, operations, customer experience, and risk, organizations must ensure that all implementations adhere to global standards for fairness, privacy, documentation, and monitoring. Without orchestration, AI ecosystems become fragmented, increasing operational risk. Therefore, effective integration merges automation with organizational alignment, creating a coherent and governed enterprise AI environment.

24.8 The Human Dimension of Integration: Change Management and Adoption

Integration is not solely a technical exercise; it is a human transformation. Employees must learn new workflows, interpret model outputs, adjust responsibilities, and adopt practices aligned with AI-augmented operations. Without cultural alignment, even the most advanced systems may be underused or misinterpreted.

Successful integration requires structured change management, which includes clear communication, role-specific training, accessible documentation, and leadership engagement. Employees must understand the purpose of AI, its benefits, its limitations, and the expectations surrounding its use. Trust and literacy are critical because when users understand the systems they interact with, adoption improves, and misuse declines. Therefore, integration rests on cultural readiness as much as technical maturity.

24.9 From Integration to Transformation: AI as an Enterprise Capability

The long-term objective of AI integration is not merely embedding models into workflows but transforming how the organization operates. When intelligence becomes a structural component of decision-making, processes evolve. Workflows gain adaptability, decision rights shift, data becomes more dynamic, and teams increasingly coordinate around predictive signals rather than static rules.

This transformation requires strategic vision. It involves aligning AI initiatives with organizational priorities, establishing governance structures that scale, and designing processes around intelligence rather than grafting intelligence onto outdated workflows. AI-driven transformation occurs when organizations treat intelligence as an enterprise capability, one that shapes strategy, culture, and operations.

24.10 Conclusion: Making Intelligence Operational

Enterprise integration marks the point where AI transitions from potential to impact. It connects predictive and generative capabilities to the systems, workflows, and decisions that define organizational performance. Successful integration requires technical rigor, architectural coherence, thoughtful interface design, governance alignment, and cultural readiness.

CHAPTER 25

Automation, Agents, and AI-Augmented Operations

25.1 Introduction: When Intelligence Becomes an Operator

Modern enterprises operate within environments defined by complexity, speed, and continual change. Supply chains shift rapidly, customer expectations evolve, and decisions must be made under uncertainty at scale. Historically, organizations relied on rules-based automation, deterministic scripts and workflows capable of executing predictable tasks, to manage operational demand. The emergence of artificial intelligence fundamentally changes this paradigm. Automation is no longer limited to mechanical execution. Instead, it becomes adaptive, context-aware, and capable of informed decision-making.

AI-driven automation and intelligent agents now function as active participants in operations. These systems interpret data, recognize patterns, coordinate workflows, recommend or initiate actions and, in some cases, operate with a degree of autonomy. They augment human capability rather than simply replacing manual labor, enabling enterprises to achieve levels of responsiveness, precision, and scale unattainable through traditional methods. As intelligence becomes operational, organizations shift from viewing automation as a cost-reduction mechanism to recognizing it as a strategic asset that strengthens resilience, agility, and innovation.

This chapter explores how intelligent agents and automation systems transform daily operations by executing tasks, coordinating workflows, and adapting in real time. It examines how organizations transition from human-led execution to AI-augmented operational ecosystems. It also highlights how this shift reshapes productivity, oversight, and operational risk.

25.2 From Rules-Based Automation to Adaptive Intelligence

Rules-based automation depends on predictability, where processes must be stable, well-defined, and expressible through fixed logic. When conditions shift or inputs deviate, traditional systems often break down. AI changes this paradigm by introducing adaptive decision-making into the automation layer, allowing systems to interpret data, infer context, and choose actions dynamically rather than follow rigid instructions. This shift expands the types of processes automation can meaningfully handle. For example, customer service systems can interpret nuanced inquiries, fraud detection models evolve with adversarial behavior, and forecasting engines adjust decisions based on real-time signals. The defining transformation is flexibility: AI allows automation to operate effectively amid uncertainty, variability, and complexity.

Realizing this capability, however, requires intentional design. Adaptive automation must be grounded in high-quality data, sound governance, and clear oversight to ensure that decisions remain aligned with organizational goals. Boundaries between automated actions and human judgment must be explicit to prevent ambiguity in accountability or risk management. When implemented responsibly, adaptive intelligence elevates automation from a narrow technical tool to a strategic asset, which enhances responsiveness, resilience, and competitive differentiation.

25.3 The Rise of Intelligent Agents

Intelligent agents represent a major advancement in operational AI because they are designed not just to predict but to act. They perceive their environment, evaluate goals or constraints, and select behaviors that advance organizational objectives. Operating either individually or within multi-agent systems, they can collaborate, negotiate, or coordinate as needed. Unlike traditional automation, which executes fixed instructions, intelligent agents possess autonomy with the ability to choose contextually appropriate actions in dynamic conditions.

As organizations adopt these systems, agents increasingly manage workflows, monitor operations, and interact directly with customers. They can orchestrate supply chains, triage support requests, manage resource allocation, and route inquiries across distributed teams. This shift enables enterprises to operate at greater scale and responsiveness while reducing cognitive load on employees. Agent deployments foster hybrid ecosystems in which humans retain strategic direction and oversight while agents handle repetitive, time-sensitive, or high-volume tasks.

25.4 Human-in-the-Loop: Augmentation Rather Than Replacement

Despite rising autonomy in intelligent systems, human expertise remains essential for ensuring reliability and ethical alignment. Human-in-the-loop (HITL) design embeds human oversight into key stages of decision-making, allowing people to validate outputs, correct errors, and intervene when ambiguity or risk arises. This approach blends machine strengths such as speed, scale, and consistency with human strengths such as contextual understanding, ethical judgment, and empathy, creating more robust operational workflows.

In AI-augmented environments, human roles shift from manual execution to supervisory control. Employees no longer review every transaction; they focus on exceptions flagged by automated systems. Instead of coordinating workflows step-by-step, they set objectives, constraints, escalation rules, and performance thresholds for intelligent agents. This augmentation enhances productivity while preserving accountability and providing a safeguard when models encounter unfamiliar or unstable conditions. Therefore, human-machine collaboration remains a foundational principle of responsible operational design.

25.5 Decision Automation: Scaling Judgment Across the Enterprise

Decision automation embeds AI directly into business processes, allowing models to assess risk, recommend actions, and initiate workflows, often in real-time. These systems influence pricing, personalize customer interactions, guide resource allocation, and optimize supply chain routing. By distributing intelligence across routine operations, enterprises gain the ability to act quickly, consistently, and at scale, transforming decision-making from a manual bottleneck into a continuous, data-driven capability.

Realizing this potential requires clear governance and well-defined boundaries. Organizations must specify when AI may act autonomously, when it should request human approval, and when decisions should remain exclusively under human control. These guardrails are essential for maintaining trust, mitigating risk, and ensuring compliance in sensitive or regulated environments. With these structures in place, decision automation becomes a reliable mechanism for scaling judgment across the enterprise.

25.6 AI-Augmented Workflows: The New Operational Blueprint

Workflows evolve significantly when augmented by AI. Instead of following rigid sequences, they become adaptive, responding to predictions, user behavior, or environmental signals. A support workflow may reassign cases based on predicted complexity. A logistics flow may dynamically reroute shipments in response to real-time constraints. A manufacturing process may adjust production schedules according to demand forecasts.

These adaptive workflows shift organizations from static procedures to dynamic operational ecosystems. They also enable continuous improvement. As models learn from data, workflows recalibrate to become more efficient and better aligned with organizational goals. Over time, workflows emerge not solely from manual design but from the joint contributions of human insight and learned intelligence.

25.7 Agents as Orchestrators: Coordinating People, Systems, and Processes

In advanced operational environments, intelligent agents increasingly function as orchestrators that coordinate tasks, resources, and information across teams and systems. They manage interdependencies, trigger actions at the appropriate moments, and escalate exceptions when human judgment is required. Whether routing customer inquiries, synchronizing multi-step workflows, or aligning cross-functional processes, these agents bring coherence to complex operational ecosystems.

Practical applications are already widespread. In customer support, an agent may pull data from various systems, run predictive models, and present recommended actions to a representative. In supply chain management, agents can coordinate inventory flows across warehouses, carriers, and suppliers. In IT operations, they monitor infrastructure health and initiate remediation before failures escalate. This orchestration reduces bottlenecks, enables proactive coordination, and lays the groundwork for increasingly autonomous enterprise systems.

25.8 Trust, Transparency, and Control in AI-Augmented Operations

As automation becomes more intelligent and pervasive, trust and control emerge as central requirements for responsible deployment. Employees must understand how automated decisions are generated, leaders must know when and how to intervene, and customers must feel confident that automated interactions are fair, consistent, and reliable. Without this understanding, intelligent systems risk resistance, misuse, or unintended harm.

Transparency plays a critical role in establishing this trust. Systems must offer interpretable explanations, traceable decision pathways, and predictable behavior across scenarios. Complementing transparency, strong control mechanisms, including governance frameworks, monitoring tools, audit trails, and fail-safes, ensure that automated actions remain aligned with ethical standards, regulatory requirements, and organizational objectives. Together, trust and control create the foundation necessary for sustainable AI-augmented operations.

25.9 The Emerging Landscape of Autonomous Operations

As organizations progress in their AI maturity, operations evolve toward higher levels of autonomy. Initial phases focus on automating discrete tasks, but later stages enable decision-making and coordination across entire workflows. Ultimately, enterprises may deploy systems capable of acting with limited human intervention, guided by high-level goals rather than step-by-step instructions. This shift allows operations to respond more quickly and coherently to dynamic conditions.

Elements of enterprise autonomy are already visible in self-healing IT systems, dynamic optimization engines, and automated resource management tools. These capabilities reduce operational friction, enhance responsiveness to disruptions, and identify emerging opportunities faster than traditional methods. Importantly, increasing autonomy does not diminish human control. Instead, control becomes more strategic, focusing on guiding system behavior, setting constraints, and governing intelligent processes rather than executing individual tasks.

25.10 Conclusion: Intelligence as the Engine of Operational Transformation

AI-augmented operations constitute a fundamental shift in how organizations function. Automation becomes adaptive, agents emerge as collaborators, and workflows evolve into dynamic, learning systems. The enterprise transitions from static procedures to fluid, intelligence-driven operations capable of responding to changing conditions with speed and precision.

Rather than displacing people, AI elevates their roles and empowers them to focus on judgment, strategy, creativity, and complex problem-solving. Intelligence becomes the engine that drives operational transformation, enabling organizations to operate with greater agility, resilience, and sustained innovation.

CHAPTER 26

Human-Machine Collaboration and the Future of Work

26.1 Introduction: A New Relationship Between People and Intelligent Systems

Artificial intelligence is reshaping not only organizational operations but also the nature of human work, learning, and contribution. For the first time, machines are not confined to executing predefined instructions. They are capable of perceiving patterns, interpreting information, generating insights, and recommending actions. As AI becomes embedded across workflows and industries, the structure of work evolves. Tasks change, roles transform, and responsibilities shift.

The emerging paradigm is not one in which machines displace human capability, but one in which humans and intelligent systems collaborate, with each amplifying the strengths of the other. Human-machine collaboration blurs traditional boundaries between digital systems and human cognition, reshaping job design, skill requirements, and organizational culture.

This chapter examines how AI is transforming work, how collaborative relationships between people and systems develop, and how organizations can cultivate environments where human potential is enhanced rather than diminished by intelligent technologies. It also clarifies how skills, roles, and leadership models evolve in response to human-machine collaboration.

26.2 The Evolution of Work: From Automation to Augmentation

The history of work has progressed through three major transformations: mechanization of physical labor, digitization of information, and now the introduction of intelligence into the workplace. In contrast to traditional automation, which focused on replacing manual tasks through deterministic logic, contemporary AI emphasizes augmentation by supporting human performance rather than substituting for it.

AI handles repetitive, data-intensive, or time-sensitive tasks, analyzes information at scale, and identifies risks and opportunities. Human expertise becomes increasingly dedicated to supervising systems, validating recommendations, and applying contextual reasoning and ethical judgment. This shift highlights the complementary strengths of both parties. Machines provide consistency, speed, and analytical depth, while humans contribute empathy, creativity, ethical deliberation, and strategic direction. Together, they enable organizations to operate with greater agility, precision, and innovative capacity.

26.3 Redefining Roles in an AI-Infused Workplace

As AI becomes integrated into daily workflows, work roles are reshaped rather than eliminated. Routine tasks are automated, allowing employees to focus on higher-value responsibilities that require interpretation, coordination, and judgment. Decision-making becomes a hybrid process in which humans collaborate with models, using AI-generated insights as structured inputs rather than unquestioned conclusions.

Analysts become facilitators of insight rather than mere data processors. Customer service representatives evolve into orchestrators of information supported by intelligent assistants. Operations managers shift from tactical oversight to guiding agent-driven systems that manage scheduling, routing, or optimization. These transitions require new competencies, including data awareness, fluency in interpreting model outputs, and the ability to collaborate effectively with systems that reason probabilistically rather than intuitively.

26.4 AI Literacy: The Foundation of Human-Machine Collaboration

AI literacy, a shared understanding of how intelligent systems work and what their limits are, is essential for effective collaboration. AI literacy does not require all employees to become technical experts. Instead, it involves the ability to interpret predictions, distinguish between correlation and causation, recognize the meaning of uncertainty, and understand when human intervention is necessary.

Without AI literacy, employees may place excessive trust in systems, misinterpret outputs, or apply models in inappropriate contexts. With literacy, they engage critically and constructively, identifying anomalies, asking informed questions, and exercising proper oversight. AI literacy becomes a foundational competency for modern workforces, enabling responsible and effective human-machine interaction across the enterprise.

26.5 Managing Change: Navigating Fears, Expectations, and Resistance

The introduction of AI often generates apprehension, due to fear of job displacement, concern about system reliability, or uncertainty about changing responsibilities. These reactions reflect legitimate anxieties regarding technological transformation. Effective change management acknowledges these concerns and addresses them proactively through transparency, communication, and inclusivity.

Organizations must clearly articulate the purpose of AI adoption, its benefits, and its implications for daily work. Employees should be engaged early, provided with opportunities to learn, and supported through transitions. Incremental deployment allows trust to build gradually. When employees feel included and valued, they view AI as a partner rather than a threat. Thus, change management becomes as much a social and psychological process as a technical one.

26.6 Designing Human-Centered AI Systems

AI systems function best when developed through human-centered design. Such systems prioritize usability, clarity, and trustworthiness. They present information in interpretable formats, communicate uncertainty, and provide explanations that illuminate underlying reasoning. They are designed to support human cognition rather than overwhelm it.

Human-centered design ensures that users retain meaningful control and that AI augments rather than replaces human judgment in scenarios requiring context, empathy, or ethical reasoning. Recognizing that intelligence is both computational and relational, human-centered AI seeks to deepen collaboration by aligning design choices with human needs, capabilities, and limitations.

26.7 The Role of Human Oversight in an AI-Augmented Organization

Even in environments where AI performs substantial operational functions, human oversight remains indispensable. Oversight provides safeguards against model drift, unexpected behavior, and ethical risks. Humans validate outputs, adjudicate ambiguous cases, and make decisions that require contextual or moral reasoning.

Oversight also mitigates automation bias, ensuring that human judgment remains active rather than passive. In regulated sectors, including finance, healthcare, and public services, human oversight is not only prudent but required. Therefore, the future of work involves both collaboration and governance, ensuring alignment between automated behaviors and organizational values.

26.8 The Skill Transformation: What Workers Need in the Age of AI

As AI reshapes organizational roles, required skills evolve as well. Technical literacy becomes increasingly valuable, but so do uniquely human capabilities such as critical thinking, creative problem-solving, ethical analysis, communication, and interdisciplinary collaboration. Organizations that invest in reskilling and upskilling prepare their workforce to thrive alongside intelligent systems.

Because AI technologies evolve rapidly, learning must be continuous rather than episodic. Effective workforce development embeds learning into everyday practice, encouraging employees to adapt, explore new tools, and participate actively in shaping organizational transformation. Thus, skill transformation becomes a long-term strategic priority rather than a short-term initiative.

26.9 Collaboration, Not Competition: A New Vision of Work

The future of work is defined not by competition between humans and machines, but by collaboration. AI excels at pattern recognition, optimization, and large-scale analysis, while humans provide narrative understanding, social awareness, ethical reasoning, and strategic foresight. Together, they achieve outcomes that neither could accomplish independently.

This collaborative vision requires rethinking productivity metrics, role design, and organizational success indicators. It also requires leadership to champion an understanding of AI as an enabler of human capability. When organizations adopt this mindset, employees become more empowered, engaged, and capable, and AI becomes a catalyst for expanding rather than constraining human potential.

26.10 Conclusion: The Collaborative Enterprise of the Future

Human-machine collaboration is transforming the nature of work. It reshapes roles, redefines workflows, and expands organizational capacity. AI augments human expertise, enabling individuals to focus on higher-order thinking, creativity, and strategic decision-making. The future belongs to organizations that embrace literacy, cultivate trust, design human-centered systems, and invest continuously in people.

This collaborative model sets the stage for enterprise transformation at scale. As we move forward, the challenge is not simply to integrate AI into work, but to build work environments where human capability and machine intelligence reinforce one another, forming the foundation of a resilient, innovative, and adaptive enterprise.

CHAPTER 27

Scaling AI in Organizations: Culture, Maturity, and Governance

27.1 Introduction: Moving from Experiments to Enterprise Transformation

Most organizations begin their AI journey with isolated experiments, which include pilot projects run by enthusiastic teams, proofs of concept targeting narrow problems, or exploratory models designed to test feasibility. While such efforts generate early momentum, they do not constitute enterprise capability. Scaling AI requires a fundamental shift in mindset: from experimentation to institutionalization. It requires embedding intelligence into culture, infrastructure, governance, and strategic priorities.

The challenge of scaling is not purely technical. It involves constructing the social, structural, and managerial foundations in which AI can thrive responsibly and sustainably. Without these foundations, AI initiatives remain fragmented and fail to deliver lasting value.

This chapter examines the organizational transformations necessary for scaling AI, including reshaping culture, strengthening governance, developing maturity, and building environments in which intelligence becomes a structural competitive advantage. It also explains how leadership alignment and sustained change management determine whether scaling efforts succeed.

27.2 Why Scaling AI Is Difficult: The Organizational Reality

Many organizations underestimate the difficulty of scaling AI, often focusing on models, platforms, or specialized tools while overlooking the organizational scaffolding required to support them. Successful scaling depends on alignment across leadership, processes, data foundations, and governance. Without this alignment, even the most sophisticated initiatives struggle to progress. Problems frequently emerge when organizations attempt to expand AI efforts before defining ownership, standardizing workflows, or establishing clear decision rights. As a result, technical teams may build advanced models that never reach production because operational units cannot integrate them, and leaders may champion AI rhetorically without articulating how it advances strategic priorities.

Cultural and structural barriers further complicate scaling efforts. Employees may resist change when they fear role disruption or do not understand how AI will affect daily responsibilities. Operational teams may lack the capacity to absorb new systems, and fragmented processes can impede sustained adoption. Therefore, scaling AI requires organizational coherence. Every segment of the enterprise must understand the purpose of AI, support its integration, and participate in a shared framework that ensures technology, people, and processes advance together.

27.3 Culture as the Core Infrastructure of AI

Technology cannot compensate for a culture that does not understand, trust, or value AI. Culture shapes how employees interpret AI recommendations, how willingly they adopt intelligent tools, and whether they view automation as supportive or threatening. A strong AI culture encourages curiosity, responsible experimentation, and cross-functional collaboration.

Cultural transformation requires intentional effort. Leaders must communicate openly, demonstrate the responsible use of AI, and create an environment where employees can learn without fear. Teams must see AI not as an opaque black box but as a partner that enhances their capabilities. Psychological safety is essential because it enables employees to ask questions, raise concerns, and experiment without fear. Culture becomes the connective tissue that allows AI to integrate seamlessly into organizational identity.

27.4 Leadership and Vision: Setting the Direction for AI Maturity

Scaling AI begins with leadership that understands both the potential and the responsibility of intelligent systems. Leaders must articulate a coherent vision for how AI supports the organization's mission, advances strategic priorities, and aligns with ethical commitments.

Vision, however, must be paired with practice. When leaders use AI-driven insights to guide decisions, it signals legitimacy. When they articulate guardrails for responsible use, it establishes shared expectations. When they invest in training, infrastructure, and governance, it demonstrates commitment. Leadership defines not only the strategic direction but also the emotional tone of AI initiatives. Success depends heavily on how leaders frame the role and purpose of AI within the enterprise.

27.5 Building Organizational Maturity: From Localized Efforts to Enterprise Capability

AI maturity develops over time. Organizations typically progress in stages from small pilot projects to standardized practices and eventually to embedding AI across everyday operations and decision-making. Advancing through these stages requires deliberate investment.

Organizations must build strong data foundations, formalize documentation, standardize deployment and monitoring, and develop the human capabilities required to sustain AI, including technical expertise, literacy programs, and cross-functional collaboration. Therefore, true maturity extends beyond technology itself and is reflected in organizational readiness, operational consistency, and clear alignment with strategic goals.

27.6 Centers of Excellence and Federated Models: Structuring for Scale

Effective organizational structures enable AI to scale without chaos. Many enterprises establish AI Centers of Excellence (CoEs) as hubs of expertise, governance, and reusable best practices. CoEs support complex projects, enforce standards, and provide critical resources such as frameworks for responsible AI.

A federated model often complements the CoE. Business units retain the autonomy to build solutions tailored to their domain while operating within shared governance, infrastructure, and architectural guidelines. This hybrid structure accelerates innovation while preventing fragmentation. Central coordination supplies guardrails, while local teams contribute domain insight and creativity. Together, they form a structure capable of sustainable scaling.

27.7 Governance as the Engine of Safe and Scalable AI

Governance is not an impediment to innovation, as it is the mechanism that makes innovation sustainable. Without governance, AI development becomes unpredictable, ethically risky, and operationally fragile. With governance, organizations can innovate confidently and deploy models across critical workflows.

Effective governance defines ownership and accountability, documents model lineage, enforces monitoring requirements, and ensures compliance with fairness, transparency, and privacy standards. It establishes escalation pathways, human oversight procedures, and robust version control. Governance reduces uncertainty, enabling teams to develop, deploy, and maintain AI in consistent, auditable ways, capabilities essential as AI permeates the enterprise.

27.8 Cross-Functional Collaboration: Integrating Expertise Across Domains

Scaling AI requires interdisciplinary collaboration. Data scientists alone cannot ensure successful or responsible AI adoption. Engineers provide architectural stability, domain experts supply contextual judgment, ethicists and legal teams flag societal and regulatory risks, and product leaders integrate AI into operational reality.

Cross-functional collaboration becomes especially essential in high-stakes workflows. Domain experts identify nuances that models do not capture. Legal specialists foresee compliance issues, ethicists surface fairness concerns, and engineers ensure robustness and maintainability. In mature AI organizations, such collaboration is not an exception. It is institutionalized as standard practice.

27.9 Measuring Success: The Metrics of AI Adoption and Impact

Scaling AI requires clear metrics that extend beyond technical performance. Organizations must evaluate adoption rates, reliability, ethical alignment, operational stability, and business impact. Technical metrics alone cannot capture full enterprise value. Organizations must assess whether AI is being used appropriately and whether it produces meaningful improvements.

Success becomes evident when employees trust and rely on AI, and when leaders consistently incorporate it into decision-making. It is further reflected when governance mechanisms function smoothly and when measurable benefits appear, whether in efficiency gains, risk reduction, revenue growth, or improved customer experience. Metrics provide the clarity needed to steer AI development deliberately and responsibly.

27.10 Conclusion: Building the AI-Driven Enterprise

Scaling AI transforms organizations from experimental adopters into intelligent enterprises. It requires a deliberate blend of culture, structure, maturity, governance, and leadership. The process is iterative and adaptive, not linear. When executed thoughtfully, scaling AI increases resilience, enhances decision-making, and supports innovation that is both responsible and sustainable.

CHAPTER 28

Responsible AI: Principles for Safe, Fair, and Transparent Systems

28.1 Introduction: Responsibility as the Core of AI Maturity

As artificial intelligence becomes embedded in the daily operations of organizations, its influence extends far beyond technical systems. It shapes decisions, behaviors, opportunities, and outcomes. AI now determines who is hired, who receives credit, how patients are triaged, how resources are allocated, and how individuals experience digital services. With such pervasive impact, responsible AI shifts from an ethical aspiration to a practical requirement. Responsibility becomes the backbone of trust, compliance, operational stability, and long-term organizational legitimacy.

Responsible AI is not a single guideline or technical checklist. It is an overarching philosophy of design, development, and deployment, which acknowledges the societal implications of AI-driven decision-making. It requires organizations to ensure that intelligent systems operate fairly, transparently, accountably, securely, and in alignment with human values.

This chapter explores the foundational principles of responsible AI, including how fairness, transparency, accountability, and safety guide the design and deployment of intelligent systems. It examines how these principles are operationalized through governance structures, ethical review, and continuous oversight across the AI lifecycle.

28.2 Why Responsible AI Matters: The Human Stakes of Intelligent Systems

AI can enhance decision-making, expand access to services, and significantly improve operational efficiency. However, without safeguards, it can also cause real and lasting harm. Systems trained on biased data may perpetuate discrimination; opaque models can undermine trust; and optimization strategies focused solely on efficiency may erode fairness or inclusivity. These risks grow more severe as AI scales across industries, cultures, and geographies, influencing decisions that affect millions of people.

The consequences of irresponsible AI are tangible. These include biased hiring systems, discriminatory credit models, flawed predictive policing tools, unreliable medical triage algorithms, and platforms that generate harmful misinformation. Responsible AI seeks to prevent such outcomes by embedding ethical, social, and human considerations directly into the design and deployment process. Organizations that embrace these principles not only reduce risk but also strengthen trust, increase adoption, and build positive relationships with customers, employees, and regulators. In doing so, responsibility becomes both a moral obligation and a strategic advantage.

28.3 Fairness: Ensuring Equitable Outcomes Across Groups

Fairness is one of the most challenging components of responsible AI because it requires assessing who benefits from a system and who may be harmed. Ensuring fairness means scrutinizing whether training data reflect societal inequalities, whether certain groups are underrepresented, and whether model outputs reinforce or counteract these patterns. It demands a careful understanding of how social and demographic factors shape the data on which AI systems rely.

Achieving fairness goes far beyond applying technical debiasing methods. It requires contextual understanding of historical and cultural dynamics, continuous evaluation as conditions change, and meaningful participation from diverse stakeholders, especially in high-stakes contexts involving rights, access, or wellbeing. When fairness is prioritized, AI systems become tools for promoting equity rather than mechanisms that inadvertently deepen existing inequalities.

28.4 Transparency: Illuminating How AI Makes Decisions

Transparency is essential for trust, accountability, and effective oversight in AI systems. It ensures that users and stakeholders understand how decisions are generated, what data models rely on, and what assumptions or trade-offs guide system behavior. Transparency does not necessarily require revealing proprietary algorithms. Rather, it requires communicating decision logic clearly and accessibly so that people understand how outcomes are produced.

Explainability, the ability to articulate why a specific prediction or recommendation was made, is critical in many domains. Healthcare providers must be able to justify treatment suggestions, loan applicants deserve clarity on credit decisions, and regulators need insight into automated systems that influence public outcomes. Transparent systems help bridge the gap between algorithmic processes and human understanding, fostering confidence and supporting informed decision-making.

28.5 Accountability: Ensuring That Someone Owns the Outcomes

Accountability anchors responsible AI by ensuring that organizations remain answerable for decisions influenced or automated by models. This requires clearly defined responsibilities across model development, validation, deployment, monitoring, and governance. Without these structures, responsibility can become diffuse, a common risk in complex technical environments, resulting in gaps that undermine reliability and public trust.

Accountability also encompasses the rights of individuals affected by AI systems. People must have the ability to challenge decisions, request explanations, or seek redress when outcomes are harmful or inaccurate. Without such mechanisms, AI risks becoming an unchallengeable authority, an unacceptable scenario in ethical, legal, or democratic contexts. Embedding accountability within governance frameworks ensures that humans, not algorithms, remain ultimately responsible for outcomes.

28.6 Privacy and Data Protection: Respecting the Boundaries of Personal Information

AI systems frequently rely on sensitive personal data, making privacy and data protection central pillars of responsible AI. Ethical collection, secure storage, and clearly communicated use of personal data are essential. Principles such as consent, data minimization, purpose limitation, and secure processing must be embedded throughout the AI lifecycle. They should not be treated as regulatory checkboxes but as foundational design standards.

Privacy also intersects deeply with fairness and transparency. Overcollection of data may disproportionately affect vulnerable populations, while opaque data practices can erode trust and reduce willingness to engage with AI-powered services. Therefore, responsible AI requires more than compliance with regulations like GDPR or HIPAA. It demands a commitment to respecting autonomy, dignity, and individual rights. Data must be treated not merely as an asset but as a responsibility entrusted to the organization.

28.7 Safety and Reliability: Building Systems That Behave as Intended

Safety ensures that AI systems behave reliably, even in uncertain or dynamic environments. It requires designing models that are robust against noise, adversarial manipulation, and environmental changes. Safety also includes stress-testing models, monitoring them continuously, and implementing fail-safes that prevent harmful or unintended outcomes.

Not all tasks are suitable for automation, and not all decisions should be delegated to AI. Determining when human oversight is necessary, and ensuring it is effectively applied, is central to responsible development. Reliable systems strengthen trust, support adoption, and reduce operational and ethical risk.

28.8 Human-Centric Design: Putting People at the Center of Intelligent Systems

Human-centric design prioritizes building AI systems that empower users rather than overwhelm, confuse, or displace them. It emphasizes usability, interpretability, and meaningful human control. Interfaces should present information clearly, outputs should be actionable, and systems must communicate uncertainty, limitations, and context. The goal is to ensure that people can understand and appropriately rely on intelligent systems rather than feel alienated or mistrustful.

A human-centric approach also includes accessibility, ensuring that tools accommodate individuals with different backgrounds, abilities, and technical fluency. When technologies honor human needs and limitations, collaboration improves, adoption increases, and unintended consequences diminish. Therefore, human-centered AI is not just a design preference. It is an ethical commitment to building systems that serve people reliably and respectfully.

28.9 Governance and Process: Institutionalizing Responsible AI

Responsible AI cannot rely on individual good intentions. It must be embedded into the fabric of the organization through formal governance structures. These structures define standards, roles, processes, and oversight mechanisms that guide every phase of the AI lifecycle, from data collection and labeling to model development, deployment, monitoring, and retirement. Governance ensures that responsibility is systematic rather than ad hoc.

Effective governance requires collaboration across technical teams, legal experts, ethicists, domain specialists, and leadership. It involves documenting risks, performing audits, evaluating fairness, implementing monitoring protocols, and updating systems as conditions evolve. Strong governance transforms responsible AI from a conceptual aspiration into a disciplined, repeatable practice. It makes ethical, compliant, and trustworthy AI durable, consistent, and scalable across the enterprise.

28.10 Conclusion: Responsibility as a Strategic Imperative

Responsible AI cannot rely on individual good intentions. It must be embedded into the fabric of the organization through formal governance structures. These structures define standards, roles, processes, and oversight mechanisms across the entire AI lifecycle, from data collection and labeling to development, deployment, monitoring, and eventual retirement.

Effective governance demands collaboration among technical teams, legal experts, ethicists, domain specialists, and leadership, supported by practices such as risk documentation, audits, fairness evaluations, monitoring protocols, and continual updates as conditions change. By institutionalizing these processes, governance transforms responsible AI from an aspirational concept into a disciplined, repeatable practice that ensures ethical, compliant, and trustworthy AI is durable, consistent, and scalable across the enterprise.

CHAPTER 29

AI Strategy: Building Intelligence Competitively and Sustainably

29.1 Introduction: Strategy in the Age of Intelligent Organizations

Artificial intelligence is reshaping competitive dynamics across industries. Organizations that once differentiated themselves through scale, operational efficiency, or brand now compete on intelligence. Specifically, they compete on how quickly they can interpret their environments, anticipate change, personalize experiences, and optimize decisions. Yet intelligence cannot be achieved through tools alone. It requires strategic foresight, disciplined execution, organizational alignment, and long-term commitment.

AI strategy defines how intelligence becomes a sustained source of competitive advantage rather than a set of disconnected pilots. It aligns investments with outcomes, guides organizational design, informs governance, and shapes the human-machine collaboration that underpins daily work. A strong AI strategy clarifies not only what an organization intends to build, but why, how, and in what sequence.

This chapter explores the foundations of effective AI strategy, including how organizations build durable, scalable, and responsible competitive advantage in an intelligence-driven economy. It examines how prioritization, governance, capability development, and execution discipline shape long-term strategic success. It also highlights how strategic misalignment can quietly erode even the most advanced technical capabilities.

29.2 The Strategic Role of AI: From Optimization to Transformation

AI creates value on multiple levels, beginning with early initiatives aimed at optimization through reducing costs, accelerating processes, and minimizing errors. While these gains are valuable, organizations that limit AI to efficiency improvements miss its deeper transformative potential. Mature strategies move beyond optimization toward reinvention, by redefining products and services, reimagining workflows, enabling new business models, and creating capabilities that previously did not exist. In this strategic mode, AI becomes a force for structural change rather than a tool for incremental refinement.

AI's strategic impact emerges when it reshapes core functions. Rather than merely predicting demand, it reconstructs supply chains around dynamic forecasting. Instead of just automating customer service, it enables personalized, anticipatory engagement. It not only analyzes data but also uncovers hidden opportunities and systemic inefficiencies. When deployed with a transformative outlook, AI becomes a catalyst for innovation and differentiation, allowing organizations to compete not by doing the same things better but by doing fundamentally different things. Sustainable advantage is achieved when intelligence becomes embedded in culture, decision-making, operational processes, and strategic identity.

29.3 Defining the Organization's AI Ambition

Every organization must articulate its AI ambition and the role AI is expected to play in its future. A clear ambition guides priorities, shapes investment decisions, and sets realistic expectations for stakeholders. It determines whether AI will function as a tool for incremental improvement or serve as a mechanism for enterprise-wide transformation, and whether the emphasis will be on targeted, domain-specific applications or on building broad, cross-functional capabilities that can scale.

Ambition must reflect industry context, organizational maturity, regulatory constraints, and leadership philosophy. Digital-native companies may pursue aggressive AI integration, while organizations in highly regulated or risk-sensitive sectors may adopt more cautious, compliance-focused approaches. Whatever the stance, clarity is essential. Without a defined ambition, organizations drift and create fragmented deployments, duplicate effort, and chase trends without strategic grounding. A well-articulated ambition becomes the north star that guides coherent, long-term development of AI capabilities.

29.4 Aligning AI with Business Objectives

AI strategy succeeds only when tightly aligned with business objectives. Intelligent systems must support measurable outcomes such as improved customer satisfaction, reduced operational risk, stronger resilience, accelerated innovation, or enhanced competitive positioning. Organizations often fall into the trap of prioritizing technologically impressive projects that lack strategic relevance. Alignment ensures that resources are invested where AI can meaningfully influence performance.

Alignment requires sustained collaboration among data scientists, product leaders, domain experts, engineers, and executives. Together, these stakeholders identify use cases, assess feasibility, evaluate risk, and prioritize initiatives. The goal is not merely to deploy AI, but to deploy it in service of mission-critical outcomes. Strategic alignment converts AI from a technical asset into an organizational capability.

29.5 Building Capabilities: Infrastructure, Talent, and Governance

Building a durable AI strategy requires foundational capabilities that enable repeatable, scalable, and responsible innovation across the enterprise. These include robust technical infrastructure, such as integrated data platforms, scalable compute environments, deployment pipelines, monitoring frameworks, and integration architectures. They also require skilled talent capable of designing, evaluating, and maintaining intelligent systems. Governance frameworks serve as essential guardrails, ensuring that AI development remains ethical, secure, compliant, and aligned with organizational objectives. Together, these capabilities form the backbone that allows AI to function reliably at scale.

As organizations mature, these capabilities evolve from basic experimentation and skill-building toward standardized processes, automation, and coordinated enterprise-wide execution. Mature organizations embed AI literacy broadly across functions and treat model lifecycle management as a continuous operational discipline rather than a project-based activity. Because technologies and business needs change continuously, capabilities cannot remain static. They must be reinforced, refined, and adapted to sustain long-term strategic advantage.

29.6 Prioritizing Use Cases: Choosing What to Build First

Not all AI opportunities deliver equal value or are equally feasible, making thoughtful prioritization essential. Some use cases hold significant strategic potential but require substantial technical sophistication, while others are easier to implement but offer limited impact. Organizations must evaluate potential initiatives based on expected value, operational feasibility, data readiness, risk exposure, and alignment with broader strategic goals. Prioritization ensures that resources are directed toward opportunities that meaningfully advance both organizational capability and strategic direction.

These choices also shape the organization's learning pathway. Early projects should demonstrate clear value, build credibility, reinforce governance practices, and strengthen foundational capabilities. As AI maturity increases, organizations can expand into more complex, cross-functional, or transformative initiatives. AI adoption is cumulative, as each successful deployment expands institutional expertise, enhances infrastructure, and increases organizational readiness for the next stage of capability development.

29.7 Choosing Build vs. Buy: Strategic Considerations for AI Capability

Organizations must decide when to build AI systems internally, when to purchase commercial solutions, and when to adopt hybrid approaches. Building internally offers differentiation, customization, and greater control, but it requires significant investment in talent, infrastructure, and governance. Buying commercial tools accelerates deployment and reduces operational burden, though it may limit flexibility and introduce vendor dependence. Hybrid models, such as leveraging foundation models while adding proprietary fine-tuning, strike a balance between speed and strategic uniqueness.

An effective strategy establishes clear criteria for these decisions. Capabilities central to competitive identity or requiring deep domain expertise may justify internal development, while commoditized or non-core functions are often best addressed through off-the-shelf products. Regardless of the model chosen, build–buy decisions must reflect long-term strategic intent rather than short-term convenience. This intentionality ensures that AI capability grows in alignment with the organization's broader vision.

29.8 Creating Competitive Advantage Through Intelligence

AI becomes a durable source of competitive advantage when deployed in ways that competitors cannot easily replicate. Advantage may arise from proprietary data assets, unique modeling approaches, specialized infrastructure, or rigorous governance practices. However, the deepest differentiation emerges when intelligence is integrated holistically, shaping customer experience, operational design, strategic planning, and the way the organization reasons and coordinates internally.

While technologies themselves can be copied, organizational models, cultural practices, and human-machine collaboration patterns cannot. When intelligence becomes embedded in workflows, decision-making structures, and collective cognition, it forms a strategic capability that endures even in highly competitive markets. In this sense, AI strategy and organizational strategy become inseparable; the evolution of one inevitably drives the evolution of the other.

29.9 Long-Term Sustainability: Continuity, Adaptability, and Learning

Sustaining AI over the long term requires continuous evolution as data shifts, markets change, new regulations emerge, and risks evolve. Models must be regularly monitored, retrained, updated, or retired, while infrastructure must scale and modernize, and talent must remain current with new methods and tools. Sustainability reflects an organization's ability to maintain momentum rather than treat AI as a one-time initiative.

Ethical, regulatory, and operational resilience are equally critical. Organizations must anticipate emerging compliance requirements, respond to societal expectations, and guard against issues such as bias, drift, misuse, and systemic vulnerabilities. Long-term AI sustainability depends on the organization's ability to learn through data, failures, feedback, and the evolving technological landscape. In this way, learning becomes the engine that sustains strategic relevance over time.

29.10 Conclusion: Strategy as the Compass for Intelligent Transformation

AI strategy defines how intelligence becomes a force for competitive advantage, operational excellence, and organizational evolution. It provides direction amid rapid technological change and shifting societal expectations. A well-crafted strategy aligns ambition, infrastructure, governance, talent, culture, and business objectives. It transforms AI from a sequence of isolated experiments into a sustained, enterprise-wide capability.

CHAPTER 30

The Future of Analytics: Beyond Intelligence Toward Autonomous Systems

30.1 Introduction: Entering the Post-Analytical Era

Analytics has undergone a profound transformation, moving from descriptive reporting to predictive modeling, from dashboards to intelligent decisioning, and now from intelligence to autonomy. With advances in deep learning, generative architectures, multimodal AI, agentic systems, and self-improving models, analytics is entering a post-analytical era in which insights evolve from static outputs to dynamic, context-sensitive behaviors. Systems increasingly do more than describe or predict; they interpret, reason, collaborate, and act.

The future of analytics is therefore not defined by increasingly sophisticated reports or marginal improvements in accuracy. It is defined by systems capable of sustained reasoning, adaptive learning, operational participation, and meaningful partnership with humans.

This chapter explores the trajectory of analytics as it moves toward autonomous systems, highlighting the opportunities, risks, and paradigm shifts that will shape the next generation of intelligent enterprises. It examines how this shift transforms decision-making, organizational structure, and human–machine collaboration.

30.2 The Shift from Reporting to Reasoning

Analytics has traditionally been retrospective, focused on summarizing past events, and even predictive analytics remained largely limited to estimating future outcomes without deeper contextual understanding. The emerging frontier shifts analytics toward reasoning, the ability to interpret context, generate hypotheses, evaluate trade-offs, and propose or even execute actions. This evolution moves analytics beyond pattern recognition into a domain that supports structured deliberation and contextual decision-making.

This transformation is being driven by advances such as large language models, multimodal transformers, neuro-symbolic hybrids, and agent-based architectures capable of synthesizing information across domains. These systems allow humans to explore alternatives, assess consequences, and co-reason with intelligent tools. Future analytics will not be purely quantitative or statistical. It will be reflective, interpretive, and dynamically responsive to the complexity of real organizational environments.

30.3 Analytics as a Continuous, Autonomous Process

Traditional analytics unfolded through slow, linear cycles, collecting data, building models, and presenting results. Usually, these processes were manual and rigid. Autonomous analytics breaks from this paradigm by enabling pipelines that update continuously, models that retrain as conditions shift, and recommendations that adapt in real-time. This creates a dynamic system capable of staying aligned with rapidly evolving environments.

At the core of autonomous analytics are systems capable of self-monitoring and self-correction, which detect drift, recalibrate thresholds, resolve data conflicts, and escalate anomalies without human intervention. Achieving this requires resilient, self-healing infrastructure that can recover from pipeline failures, regenerate features, and optimize computational workloads automatically. As organizations adopt this paradigm, they will treat analytics as a living ecosystem rather than a sequence of projects, unlocking unprecedented adaptability and operational intelligence.

30.4 Agents as the New Analytical Actors

AI agents represent one of the most profound shifts in the evolution of analytics. Where traditional models act as passive tools that analysts query, agents operate as active participants. They pursue goals, gather information, evaluate conditions, plan multi-step actions, and coordinate tasks across systems. This changes analytics from a tool-assisted process into a collaborative environment in which autonomous entities contribute directly to insight generation and decision execution.

These agents now assume responsibilities once reserved for human analysts or operational leaders. They can diagnose risks, adjust workflows, and reallocate resources in real-time. In customer environments, agents anticipate needs, evaluate sentiment, personalize interactions, and resolve issues proactively. In technical settings, they oversee AI system maintenance by coordinating retraining, monitoring pipelines, and resolving data inconsistencies. By taking on these active roles, agents become true analytical actors, autonomous contributors who help organizations discover insights, test scenarios, and implement decisions, fundamentally reshaping the analytical landscape.

30.5 Multimodal Intelligence: Understanding the Full Spectrum of Data

Analytics once centered on structured data, and modern AI expanded this to include unstructured text, images, audio, and video. The next evolution is multimodal intelligence, systems that fuse all modalities simultaneously. By integrating numerical patterns with textual meaning, visual information with contextual metadata, and sensor signals with historical experience, multimodal AI delivers a far richer and more holistic analytical view.

This fusion dramatically expands what organizations can understand and act upon. Healthcare decisions may unite imaging, genomics, and clinical notes, retail strategies may merge behavioral signals with spatial analytics, and industrial operations may integrate video, telemetry, and historical maintenance records. Therefore, the future of analytics is multidimensional and immersive, shifting from narrow statistical analysis to comprehensive situational understanding across the full spectrum of organizational data.

30.6 Generative AI and Synthetic Insights

Generative AI has already reshaped how organizations create content, simulate scenarios, and produce synthetic data. The next evolution lies in generating synthetic insights, analytical conclusions derived from simulation, scenario modeling, and counterfactual exploration. These systems will allow organizations to ask probing, strategic questions such as: What if demand surges unexpectedly? What if supply networks fracture? How would new regulations ripple through operations? By exploring these hypothetical futures, organizations gain visibility into potential risks and opportunities long before they materialize.

This capacity enables enterprises to test alternatives without incurring real-world consequences, improving preparedness and resilience. Synthetic data can also supplement or replace sensitive information, supporting fairness and privacy in model development. In this way, generative AI transforms analytics from describing present realities into envisioning future possibilities, enhancing strategy, risk planning, and innovation with unprecedented sophistication.

30.7 Decision Ecosystems: Integrating Humans, AI, and Processes

The future of analytics will be shaped by decision ecosystems, interconnected environments where humans, AI systems, data pipelines, and operational processes interact fluidly. Instead of functioning as standalone analytical tools, intelligent systems become integral components of broader organizational networks. Information flows seamlessly across departments, insights circulate between analytical and operational layers, and decisions emerge from the combined contributions of models, agents, and human experts working in concert.

Within these ecosystems, AI systems generate insights, recommend actions, and run simulations, while humans provide context, strategic judgment, and ethical oversight. Agents coordinate workflows, execute tasks, and implement actions in operational systems, creating continuous feedback loops that refine future analysis. As silos dissolve, finance, marketing, operations, compliance, and risk functions engage with shared intelligence rather than isolated frameworks. The result is a dynamic decision environment in which insights are not merely delivered but actively exchanged, refined, and operationalized, enabling organizations to navigate complexity with far greater precision, responsiveness, and strategic coherence.

30.8 The Ethics of Autonomous Analytics

As analytics grows more agentic and autonomous, the ethical stakes expand significantly. Systems that act rather than merely recommend carry a greater risk of harm if deployed without proper boundaries. Therefore, organizations must define clear lines of responsibility. They must determine which decisions can be automated, which require human oversight, and which demand transparent appeal or review mechanisms. Autonomy amplifies the consequences of design choices, making careful delineation essential.

These developments intensify concerns related to fairness, transparency, privacy, accountability, and explainability. While a predictive model influences human decision-makers, an autonomous agent may bypass them entirely, raising the ethical bar. Responsible design, governance, and monitoring become essential guardrails to ensure that increasing autonomy enhances rather than undermines human wellbeing, institutional accountability, and societal expectations.

30.9 The Human Role in an Autonomous Future

Even as systems gain autonomy, humans remain central to the future of analytics. Rather than displacement, the trajectory points toward human elevation, where people define goals, set constraints, interpret nuance, provide ethical oversight, and resolve ambiguity. Humans continue to serve as stewards of organizational values and the ultimate point of accountability, ensuring automated processes reflect human context and judgment.

In this evolving landscape, human roles shift toward strategic decision-making, oversight, creativity, and governance. Workers become orchestrators rather than executors, guiding analytical agents, shaping priorities, and aligning autonomous intelligence with organizational purpose. The highest form of organizational intelligence emerges not from machine substitution but from human-machine symbiosis, where each augments the strengths of the other.

30.10 Conclusion: Beyond Analytics, Toward Autonomous Understanding

The future of analytics reaches far beyond better dashboards or higher predictive accuracy. It envisions systems that reason, act, adapt, and collaborate by integrating multimodal understanding, autonomous decision-making, synthetic simulation, and continuous learning into everyday operations. Even as autonomy expands, the core purpose of analytics remains constant: to deepen understanding of customers, operations, risks, opportunities, and the broader environment. Autonomous analytics amplifies this mission by extending human perception and cognitive reach, enabling organizations to grasp complexity in ways previously unattainable.

As this new era unfolds, success will hinge on responsible design, strategic foresight, and a steadfast commitment to aligning intelligence with human values. Autonomous analytics is not merely a technological evolution but a philosophical shift, reshaping how organizations interpret information, make decisions, and influence the world around them.

APPENDIX A

Modern AI Platforms and Ecosystem

Artificial intelligence development no longer revolves around isolated toolsets or standalone environments. Instead, a few dominant platform ecosystems have emerged, each offering unified pipelines for data preparation, model training, deployment, monitoring, governance, and large-scale automation. These platforms shape the global AI landscape and reflect the broader trend toward composable, cloud-agnostic, multimodal, and safety-aware intelligence.

What follows is a narrative overview of the leading platforms, their philosophies, strengths, and the roles they play in today's AI ecosystem.

Google Cloud – Vertex AI & Gemini Ecosystem

Google's AI platform represents the convergence of deep research, advanced model training infrastructure, and large-scale orchestration. Vertex AI emphasizes end-to-end lifecycle support: from data ingestion and labeling to training, fine-tuning, vector search, and real-time deployment. Its integration with Gemini models introduces multimodal capabilities that span text, vision, audio, and code, enabling organizations to unify analytics, machine learning, and generative AI within a single architecture.

Google excels in retrieval-augmented generation (RAG) pipelines, scalable embedding systems, and automated model evaluation. The platform is designed for organizations seeking cutting-edge experimentation with strong MLOps standards. Vertex AI's philosophy centers on enabling openness, supporting custom models, third-party models, and a broad ecosystem of developer tools while grounding outputs in trusted data sources.

Microsoft Azure – Azure AI Studio & Copilot Stack

Azure's AI ecosystem reflects Microsoft's core strength: enterprise readiness, security, governance, and seamless integration into business workflows. Azure AI Studio unifies model catalogues, prompt orchestration, evaluation tools, and content filters with ease of deployment across Azure's global cloud.

The Copilot architecture integrates generative AI into daily work across documents, meetings, code editing, and operations, creating a distributed AI layer throughout the Microsoft suite. Azure prioritizes responsible AI through detailed policy controls, monitoring dashboards, alignment tools, and embedded safety guardrails.

Organizations choose Azure when they require tight integration with corporate identity, compliance frameworks, and multi-layered governance. Its design philosophy centers on operational reliability and human-AI collaboration across enterprise systems.

Amazon Web Services – AWS Bedrock & SageMaker

AWS continues to dominate infrastructure-driven AI. With the introduction of AWS Bedrock, Amazon designed a neutral model-hosting environment offering models from Anthropic, AI21, Amazon's Titan family, Cohere, Meta, Stability, and others. This breadth makes Bedrock the preferred choice for organizations that want flexibility rather than vendor lock-in.

SageMaker remains a foundational platform for traditional machine learning, which is reliable, scalable, and deeply integrated with storage, compute, and streaming services. Together, Bedrock and SageMaker form a dual ecosystem: generative AI combined with classical machine learning, all supported by AWS's robust monitoring, logging, and security infrastructure.

AWS's philosophy emphasizes versatility, scale, and modularity, allowing organizations to build precisely the AI pipelines they need without forcing a single architectural model.

NVIDIA – NIM, DGX, and the Accelerated AI Stack

NVIDIA sits at the center of modern AI infrastructure. Its GPUs power nearly every major AI breakthrough, but its role now transcends hardware. NVIDIA NIM (Neural Infrastructure for Models) offers optimized, ready-to-deploy containers for LLMs, vision models, diffusion models, and multi-modal systems. DGX Cloud provides elastic, high-performance clusters suited for foundation model training.

NVIDIA's ecosystem supports both training and inference at scale, emphasizing speed, efficiency, and low-latency deployment. It is the platform of choice for organizations building custom models or deploying multimodal architectures in production. NVIDIA's philosophy is simple: enable the world's most computationally intensive AI systems to operate at peak performance and reliability.

IBM – watsonx Platform

IBM's watsonx ecosystem focuses on trust, governance, evaluation, and risk management, reflecting IBM's long-standing orientation toward regulated industries, government institutions, and mission-critical environments. The platform integrates three major pillars: watsonx.ai for model training and fine-tuning, watsonx.data for scalable data storage, and watsonx.governance for transparency, fairness, and compliance.

IBM emphasizes explainability, model tracking, and ethical oversight. Organizations choose IBM when they require AI systems that must be auditable, traceable, and aligned with strict regulatory frameworks. IBM's design philosophy is to merge innovation with rigor, AI that is powerful but always accountable.

Oracle – Oracle AI Infrastructure and OCI Vision/Language Models

Oracle approaches AI from the vantage point of enterprise databases and mission-critical transactional systems. The Oracle Cloud Infrastructure (OCI) emphasizes high-performance GPUs, optimized model training, and shared responsibility for data governance. Oracle's pretrained vision and language models are integrated directly into business workflows, ERP environments, and data platforms.

The company's strength lies in delivering AI where structured, mission-critical data lives, including finance, supply chains, manufacturing, and core enterprise operations. Oracle's philosophy centers on reliability, security, and deeply integrated enterprise intelligence.

SAS – Viya AI & Analytics Platform

SAS remains a major force in industries that require validated, auditable, statistically grounded AI, such as finance, healthcare, government, and scientific domains. The SAS Viya platform combines classical statistical modeling with modern machine learning and responsible AI components.

Viya emphasizes transparency, stability, and detailed lineage, making it suitable for organizations that require defensible models and rigorous compliance documentation. Its philosophy reflects discipline: predictable, well-validated AI aligned with regulatory expectations.

TensorFlow / PyTorch – Open-Source AI Foundations

Although not platforms in the cloud-provider sense, TensorFlow and PyTorch remain foundational components of global AI innovation. PyTorch dominates research and modern deep learning development, while TensorFlow powers scalable production systems with strong tooling and ecosystem depth.

Both frameworks contribute to a global open-source foundation upon which cloud providers, startups, and enterprise systems build their AI capabilities. Their philosophy is openness, grounded in community-driven innovation and shared frameworks for advancing the state of the art.

Conclusion

The modern AI platform landscape reflects a shift toward:

- Unified pipelines
- Multimodal intelligence
- Responsible governance
- Flexible model hosting
- Scalable infrastructure

While the specific vendors and tools evolve rapidly, the underlying roles they play in compute, orchestration, alignment, governance, and safe deployment remain foundational to AI maturity.

APPENDIX B

Glossary of AI Terms

Artificial intelligence introduces a vocabulary that blends mathematics, computer science, ethics, governance, and organizational strategy. Many of these terms describe technologies that evolve rapidly, while others refer to foundational concepts that remain consistent across generations of AI systems. The purpose of this glossary is not to offer rigid, technical definitions but to provide clear, intuitive explanations that help readers understand how these terms fit within the broader context of AI infrastructure, risks, and responsible deployment.

Artificial Intelligence

Artificial intelligence (AI) refers to systems capable of performing tasks that normally require human reasoning, perception, or decision-making. Rather than executing fixed instructions, AI systems interpret data, recognize patterns, and generalize to new situations. Modern AI spans predictive models, generative systems, reasoning engines, and autonomous agents that learn from experience and adapt to changing conditions.

Machine Learning

Machine learning is a subset of AI in which systems learn relationships from data rather than relying on predefined rules. A machine learning model analyzes examples, identifies patterns, and uses those patterns to make predictions or classifications. Machine learning is the analytical backbone of most modern AI systems and forms the foundation upon which deep learning, generative models, and autonomous intelligence are built.

Deep Learning

Deep learning uses multi-layer neural networks to model complex relationships in data. These networks learn hierarchical representations, where early layers detect simple features and later layers build abstract understanding. Deep learning enables breakthroughs in image recognition, language understanding, voice processing, and generative modeling. Its power comes from its flexibility, though this same complexity makes its behavior more difficult to explain.

Neural Networks

A neural network is a computational structure inspired by the interconnected neurons of the human brain. Each neuron receives inputs, applies mathematical transformations, and passes the signal forward. Through training, the network adjusts the strength of these connections, gradually improving its ability to map inputs to outputs. Neural networks serve as the core architecture for deep learning models.

Training Data

Training data is the collection of examples used to teach an AI system how to identify patterns. The quality, representativeness, and integrity of training data directly determine the system's performance. When training data is biased, incomplete, poisoned, or unrepresentative, the model learns distorted patterns that may lead to harmful or inaccurate outcomes.

Model Drift

Model drift occurs when an AI system's performance degrades because real-world conditions change in ways that differ from the data it was trained on. As populations evolve, behaviors shift, or environments change, the model's learned representations become outdated. Drift is inevitable in dynamic environments and must be monitored continuously to preserve accuracy and fairness.

Hallucination

Hallucination describes the tendency of generative models to produce content that is fluent, confident, and entirely incorrect. Rather than indicating deception, hallucination reflects the model's reliance on statistical patterns rather than grounded truth. Managing hallucinations requires grounding, retrieval systems, evaluation, and human oversight.

Bias

Bias in AI refers to systematic patterns of unfairness arising from biased data, flawed assumptions, or unintended correlations. Bias can cause models to favor or disadvantage individuals or groups, often in subtle ways. Addressing bias requires careful data curation, fairness testing, ethical oversight, and continuous monitoring.

Explainability

Explainability is the ability to understand how and why an AI system produces its predictions. Explainable systems provide transparency into the factors influencing their decisions, allowing humans to interpret outcomes, audit behavior, and identify potential errors. As models grow more complex, explainability becomes essential for building trust and ensuring accountability.

Alignment

Alignment refers to the degree to which an AI system's behavior matches human values, intentions, and expectations. Misalignment can occur when AI interprets objectives too literally or optimizes for goals in harmful ways. Alignment requires iterative training, human feedback, and ethical oversight to ensure that AI behaves responsibly.

Adversarial Attack

An adversarial attack manipulates an AI system by introducing inputs specifically crafted to confuse or mislead it. These attacks exploit vulnerabilities in model architecture, causing the system to misclassify data or produce incorrect outputs. Adversarial attacks reveal the fragility of intelligence built on statistical approximations.

Data Poisoning

Data poisoning occurs when attackers corrupt the training data to manipulate model behavior. Poisoned data may appear legitimate but carries hidden patterns that bias or compromise the model. Once ingested during training, the corruption becomes embedded in the system's logic, making poisoning one of the most dangerous threats to AI integrity.

Generative AI (GenAI)

Generative AI models create new content such as text, images, audio, code, and multimodal experiences by learning the underlying structure of data. These models do not simply retrieve information. Instead, they generate new outputs based on patterns learned during training. While generative AI offers immense creative potential, it also raises concerns about misinformation, authenticity, and ethical boundaries.

Large Language Model (LLM)

An LLM is a generative model trained on vast amounts of text to understand and produce human-like language. LLMs learn grammar, knowledge, reasoning patterns, and context by processing billions of tokens. They excel at summarization, conversation, content generation, and problem-solving, but are prone to hallucinations when insufficiently grounded.

Retrieval-Augmented Generation (RAG)

RAG combines generative models with external knowledge sources. When a user submits a query, the system retrieves relevant documents and provides them to the model as context. This approach grounds responses in verified references, improving accuracy and reducing hallucinations. RAG architectures are essential for enterprise use of AI.

MLOps / LLMOps

MLOps and LLMOps are operational frameworks that oversee the lifecycle of machine learning and language models. They include versioning, deployment, monitoring, retraining, governance, and risk management. These frameworks ensure that AI systems remain stable, aligned, and compliant in production environments.

AI Governance

AI governance is the collection of policies, processes, oversight mechanisms, and cultural norms that ensure AI is developed and deployed responsibly. It includes documentation, monitoring, auditing, risk assessment, human oversight, ethical review, and regulatory compliance. Governance is the structural foundation of responsible intelligence.

Autonomous Systems

Autonomous systems are AI-driven entities capable of making decisions and taking actions without direct human supervision. Examples include autonomous vehicles, robotics, drones, and multi-agent systems. While autonomy introduces efficiency, it also raises profound safety and alignment challenges requiring rigorous oversight.

AI Risk

AI risk encompasses the potential for harm arising from model inaccuracies, bias, hallucinations, adversarial attacks, poisoning, drift, misuse, governance failures, or systemic dependencies. Responsible AI requires identifying, assessing, and mitigating these risks across the entire lifecycle.

INDEX

F

G

H

I

L

S

T

U

V

W

About the Author

Arshad Khan is an analytics and artificial intelligence professional with over four decades of experience spanning consulting, corporate leadership, and education. He has worked with or advised leading organizations including Accenture, PwC, Deloitte, Chevron, Bose Corporation, Chrysler, Genentech, PepsiCo, Hitachi-America, and numerous other Fortune 500 companies.

Mr. Khan served as Program Manager for a major AI initiative in Silicon Valley, where he led efforts to design and implement advanced data and machine learning solutions. This hands-on leadership role deepened his expertise in AI strategy, deployment, and governance.

A prolific author, he has written twenty books, including seven on business intelligence and analytics and three on artificial intelligence. His publications are widely used by professionals and students for their clarity, practical focus, and depth of insight.

As an adjunct professor for more than 35 years, Mr. Khan has taught data visualization, analytics, and artificial intelligence at ten universities, including Golden Gate University, Santa Clara University, and the University of California extensions at Berkeley, Santa Cruz, and San Diego.

He holds a graduate degree in chemical engineering and an MBA, combining technical rigor with business acumen. Throughout his career, Mr. Khan has bridged the worlds of technology, education, and applied analytics–helping organizations and individuals harness the power of data and AI responsibly and effectively.

Made in United States
Orlando, FL
04 November 2026

90201008R00128